Y0-IWC-185

Springer Theses

Recognizing Outstanding Ph.D. Research

Aims and Scope

The series "Springer Theses" brings together a selection of the very best Ph.D. theses from around the world and across the physical sciences. Nominated and endorsed by two recognized specialists, each published volume has been selected for its scientific excellence and the high impact of its contents for the pertinent field of research. For greater accessibility to non-specialists, the published versions include an extended introduction, as well as a foreword by the student's supervisor explaining the special relevance of the work for the field. As a whole, the series will provide a valuable resource both for newcomers to the research fields described, and for other scientists seeking detailed background information on special questions. Finally, it provides an accredited documentation of the valuable contributions made by today's younger generation of scientists.

Theses are accepted into the series by invited nomination only and must fulfill all of the following criteria

- They must be written in good English.
- The topic should fall within the confines of Chemistry, Physics, Earth Sciences, Engineering and related interdisciplinary fields such as Materials, Nanoscience, Chemical Engineering, Complex Systems and Biophysics.
- The work reported in the thesis must represent a significant scientific advance.
- If the thesis includes previously published material, permission to reproduce this must be gained from the respective copyright holder.
- They must have been examined and passed during the 12 months prior to nomination.
- Each thesis should include a foreword by the supervisor outlining the significance of its content.
- The theses should have a clearly defined structure including an introduction accessible to scientists not expert in that particular field.

More information about this series at http://www.springer.com/series/8790

$129.00
6/9/15

Laura Bianca Bethke

Exploring the Early Universe with Gravitational Waves

Doctoral Thesis accepted by
Imperial College London, UK

Springer

Author
Dr. Laura Bianca Bethke
Blackett Laboratory, Theoretical Physics Group
Imperial College London
London
UK

Supervisor
Prof. Arttu Rajantie
Department of Theoretical Physics
Imperial College London
London
UK

ISSN 2190-5053 ISSN 2190-5061 (electronic)
Springer Theses
ISBN 978-3-319-17448-8 ISBN 978-3-319-17449-5 (eBook)
DOI 10.1007/978-3-319-17449-5

Library of Congress Control Number: 2015938738

Springer Cham Heidelberg New York Dordrecht London
© Springer International Publishing Switzerland 2015
This work is subject to copyright. All rights are reserved by the Publisher, whether the whole or part of the material is concerned, specifically the rights of translation, reprinting, reuse of illustrations, recitation, broadcasting, reproduction on microfilms or in any other physical way, and transmission or information storage and retrieval, electronic adaptation, computer software, or by similar or dissimilar methodology now known or hereafter developed.
The use of general descriptive names, registered names, trademarks, service marks, etc. in this publication does not imply, even in the absence of a specific statement, that such names are exempt from the relevant protective laws and regulations and therefore free for general use.
The publisher, the authors and the editors are safe to assume that the advice and information in this book are believed to be true and accurate at the date of publication. Neither the publisher nor the authors or the editors give a warranty, express or implied, with respect to the material contained herein or for any errors or omissions that may have been made.

Printed on acid-free paper

Springer International Publishing AG Switzerland is part of Springer Science+Business Media (www.springer.com)

Supervisor's Foreword

So far, progress in our understanding of the early Universe has been driven by observations using electromagnetic waves, such as visible light and, most importantly today, the Cosmic Microwave Background (CMB) radiation which was emitted when the Universe was 300, 000 years old. However, this approach is restricted because before that time, the Universe was opaque to electromagnetic waves. Therefore, any signals from earlier times can only be observed indirectly through their imprinted signatures on the CMB, galaxy distribution or other properties of the current Universe. This is a serious limitation, because many important questions in cosmology require information about the events during the first nanosecond after the Big Bang.

There is still a lot of scope to exploit the whole electromagnetic spectrum better, but in the long term, gravitational waves are likely to emerge as an increasingly important new source of information about the early Universe. Gravitational waves are vibrations of spacetime itself, predicted by Einstein's theory of General Relativity. Like electromagnetic waves, they travel at the speed of light, but they interact very weakly with matter. Therefore the Universe has always been transparent to them, and they can give a direct view to its very first moments. However, this also makes it much harder to exploit them in practice. Indeed, gravitational waves have not even been directly detected yet, although there is strong theoretical and indirect evidence for their existence. One could therefore say that in gravitational wave astronomy, we are still in the pre-Galilean era. With electromagnetic waves, we can now make observations that Galileo could not have imagined, and similarly, technological progress will eventually make serious gravitational wave cosmology possible.

In this thesis Dr. Laura Bethke lays down some important pieces of groundwork for the future progress of gravitational wave cosmology. The thesis begins with a comprehensive and thorough introduction to the relevant aspects of inflationary cosmology and the physics of gravitational waves, followed by original work on two separate but related topics: Gravitational waves produced during inflation in quantum gravity models, and gravitational waves produced at the end of inflation.

The work in Section 2 was carried out under the supervision of Professor João Magueijo during the first year of Dr. Bethke's Ph.D. research. It uses the Ashtekar variables of loop quantum gravity to calculate the spectrum of tensor modes, i.e., gravitational waves, produced during inflation, an early period of accelerating expansion of the Universe. It is shown that in this case the tensor perturbations are chiral, giving rise to a non-zero correlation between the CMB temperature and B-mode polarisation, in contrast with conventional theories. This correlation is potentially observable with future CMB experiments, and if detected, it will provide an observational probe of quantum gravity. In spring 2014, there was already considerable excitement about a possible detection of primordial B-mode polarisation by the BICEP2 experiment. Closer analysis showed that the signal was not of cosmological origin, but there are many other existing or planned experiments searching for B-modes, and therefore a real detection is possible even in the near future.

Section 3 switches the attention to the dynamics at the end of inflation. When inflation ended, the energy density that was driving the accelerating expansion was rapidly released and heated the Universe up to a very high temperature. This process, known as reheating, depends sensitively on the microscopic physics responsible for inflation, and therefore it provides a way to distinguish between different theories. The spectrum of gravitational waves produced during reheating has been identified as a promising observable probe of this era. Based on work carried out in collaboration with Daniel Figueroa (the University of Geneva) under my supervision, Dr. Bethke demonstrates that the gravitational wave background arising from reheating is anisotropic, meaning that the intensity of gravitational waves will be different in different directions on the sky. The relative amplitude of this anisotropy is high, making this effect potentially observable with future gravitational wave experiments. This would provide a new, rich source of information about fundamental physics and the microscopic origin of inflation.

The technological challenges in gravitational wave cosmology are massive. Current experiments such as Advanced LIGO are mainly aiming for a simple detection of gravitational waves. Whilst there are plans for more ambitious future experiments such as a space-based gravitational wave observatory (due to be launched by the European Space Agency in 2034), these will be mainly sensitive to lower frequencies than predicted by typical theories of inflation. New technology for detecting and measuring gravitational waves at higher frequencies will therefore be needed, but there are some promising ideas being explored. Eventually these practical obstacles will be overcome, and then the theoretical work presented in this thesis will reach its full importance.

London
February 2015

Prof. Arttu Rajantie

Abstract

In this thesis, I will discuss two separate topics which are related to gravitational wave production in the early universe.

The first part will focus on the tensor power spectrum from inflation, derived using the Ashtekar variables of loop quantum gravity. This formalism is different from the ordinary approach in that it uses a complex connection as the central gravitational variable instead of the metric. Although the choice of variables should not affect any classical results, it becomes vital when considering quantum mechanical quantities like vacuum fluctuations. We will find that in this formalism, the tensor power spectrum is chiral, which would lead to a non-zero *TB* correlator in the CMB. Obtaining the full *TB* power spectrum would enable us to probe this chirality and provide clues about the nature of gravity.

In the second part, I will consider gravitational waves produced from massless preheating, during which the inflation transfers energy to a scalar field χ. If χ is light, it acquires a scale invariant spectrum of perturbations from inflation. At the time of preheating, the field will therefore have fluctuations on superhorizon scales and take a different value in different parts of the observable universe. I will study GW production for different initial values of χ numerically using 3D lattice simulations. The GW amplitude strongly depends on this initial value, leading to a GW background that is anisotropic today, with relative fluctuations of order 1 %. In general, anisotropies will occur in any model of preheating with a light scalar field, and the characteristics should strongly depend on the model parameters. If a GW background from preheating was measured in the future, it would provide a novel way to distinguish between different inflationary scenarios.

Acknowledgments

During my Ph.D. and the completion of this thesis I have received invaluable assistance and advice. First and foremost, I want to thank my supervisor Arttu Rajantie for being so approachable and helpful, and for taking me on as a student a year into my Ph.D. I would also like to thank Dani Figueroa, who has been a fantastic collaborator and taught me a lot about physics.

There have been many others who have helped me throughout my time at Imperial. I want to thank David Weir for assisting me with computational problems with his great technical know-how; Chris Isham for his advice on my first project; Fay Dowker for her help during my change of supervisor; and Mia Hughes for moral and technical support at all times. I am also very grateful to my family for taking interest in my work and helping me through stressful times. Last but not least, thanks to all my fellow students for making the time of my Ph.D. so enjoyable!

Contents

Abbreviations

ADM	Arnowitt Deser Misner
BBO	Big Bang Observer
BICEP	Background Imaging of Cosmic Extragalactic Polarization
BOOMERanG	Balloon Observations of Millimetric Extragalactic Radiation and Geophysics
CMB	Cosmic Microwave Background
COBE	Cosmic Background Explorer
CR	Commutation Relation
eLISA	Evolved Laser Interferometer Space Antenna
EM	Electromagnetism
EW	Electroweak
FRW	Friedmann Robertson Walker
GR	General Relativity
GUT	Grand Unified Theories
GW	Gravitational Wave
IR	Infrared
LHC	Large Hadron Collider
LHS	Left Hand Side
LIGO	Laser Interferometer Gravitational Wave Observatory
LQG	Loop Quantum Gravity
QFT	Quantum Field Theory
RHS	Right-Hand Side
UV	Ultraviolet
WMAP	Wilkinson Microwave Anisotropy Probe

Chapter 1
Introduction

Do not look at stars as bright spots only. Try to take in the vastness of the universe.

—Maria Mitchell

Cosmology is the study of the evolution of our universe, from the Big Bang to the formation of galaxies. Out of its 13.6 billion year history, we understand all but the first fraction of a second fairly well: As the universe expands and cools, it undergoes a series of phase transitions, most notably electroweak symmetry breaking when the weak gauge bosons acquire mass; when it is a few seconds old, the first elements are formed during nucleosynthesis; after several tens of thousands of years matter rather than radiation comes to dominate the energy density; 380,000 years in the Cosmic Microwave Background (CMB) is released; and all the while structure has been forming due to the presence of small perturbations in the initial density distribution, culminating in the formation of large structures like galaxies [1].

There are still many unsolved problems surrounding this vast era, like the origin of baryon asymmetry [2], the nature of dark matter [3] and dark energy [4, 5]. However, most mysterious of all are the first few instants after the Big Bang, during which the energy density was so high that we can never hope to probe such scales directly. Instead, we need to understand how the universe we observe today could have originated, and identify suitable models for this early period, which makes it a fascinating playground for theoretical cosmologists.

Nowadays, the most accepted and widely popularised theory of the universe when it was a tiny fraction of a second old is inflation, a period of rapid expansion. This is driven by an as-yet unidentified source referred to as the inflaton, usually taken to be one (or several) scalar fields or some scalar condensate. The idea of an inflationary phase in the early universe was proposed independently by several physicists between 1979 and 1980 [6–9]. Such a mechanism solves many mysteries that have plagued cosmologists in the past, like the observed large scale homogeneity and isotropy of our universe. Most importantly, it provides a seed for all the structure we observe today. Proving whether inflation indeed occurred, and finding out exactly

© Springer International Publishing Switzerland 2015

L.B. Bethke, *Exploring the Early Universe with Gravitational Waves*,
Springer Theses, DOI 10.1007/978-3-319-17449-5_1

how, could therefore shine light onto the age old question of how we came into existence. Furthermore, it could also provide one of very few ways to link observations with theories of quantum gravity, as at no other time in our history will energies have been high enough to probe the Planck scale, at which some new physics must leave its mark.

To test the validity of inflation, cosmologists primarily resort to the analysis of the CMB, the "leftover" radiation from the Big Bang. It was first discovered in 1964 [10] when Arno Penzias and Robert Wilson, using an antenna built to measure radio waves, detected a uniform background of microwave radiation with a temperature of 2.7K and a near-perfect blackbody spectrum. It was soon realised [11] that this background corresponded to radiation that had cooled down due to the expansion of the universe and was emitted very early on, at the recombination redshift $z = 1100$, when the temperature was about a thousand times higher. Even earlier, the photons were tightly coupled to the baryons, but at the time of recombination neutral atoms formed and photons could eventually decouple from the plasma and free-stream to us observing them today.

It was not until the COBE satellite [12] measured the background more precisely in 1992 (with the results vastly improved upon by later experiments like BOOMERanG [13], WMAP [14] and Planck [15]), that cosmologists were able to observe the tiny fluctuations (one part in a hundred thousand) in the CMB temperature. These fluctuations must have been laid down during the time of inflation, and analysing them could indirectly provide information about the conditions right after the Big Bang. Cosmologists try to understand which models of inflation are viable by studying the statistics of these photon perturbations. Despite the constraints from the most recent data [16] there is still a vast number of scenarios that are compatible with the universe we observe. If the recent detection of B-mode polarization of the CMB by the BICEP2 collaboration [17] turns out to be of primordial origin,[1] it will enable us to constrain the parameter space further, especially when confirmed by other experiments and complemented with the polarization data from Planck. Still, it is unlikely that we will be able to single out a model of inflation using CMB measurements alone.

Gravitational waves (GWs) could provide a new way of understanding the conditions in the very early universe. Predicted by general relativity [19], they should arise in a number of cosmological and astrophysical settings, particularly during inflation and phase transitions shortly after. Whereas scalar perturbations of the metric during inflation are the source of the density perturbations (and therefore structure), GWs correspond to tensor perturbations [20]. These tensor modes have an impact on the polarization of the CMB, but may also potentially be measured directly. Additionally, GWs are produced from classical field inhomogeneities [21] by non-equilibrium phenomena after inflation, for example during preheating.

[1]Note that at the time of writing this thesis, there was still no consensus on whether the BICEP2 team had actually detected gravitational waves. By now it is clear that their data cannot give us conclusive evidence as their signal was dominated by galactic dust [18].

So far, due to their low amplitude (a consequence of the weakness of the gravitational force), no GWs have been directly detected yet. The first indirect evidence was provided in 1974 from the energy loss of the Hulse-Taylor binary pulsar [22]. Two major experiments will attempt to directly measure GWs from astrophysical sources in the near future, Advanced LIGO [23] (operational from 2015) and eLISA [24] (launching in 2032). Neither of these will, most likely, be able to detect the cosmological signals that would tell us about the nature of the early universe, however we can be hopeful that future generations of detectors might be up to the task.

Cosmology is a vast field and there are a number of good textbooks on the subject, of which [1, 20, 25, 26] have been important in providing the physics background of this thesis. In this introduction, I will give a general overview of Cosmology, focussing on the aspects that are important for the work presented in Chaps. 2 and 3.

In Sect. 1.1, I will introduce key concepts of general relativity and describe how the homogeneous universe can be described using the Friedmann equations. The theory of inflation is the topic of Sect. 1.2, and I will explain its classical as well as its quantum aspects. Section 1.3 provides the link between inflation and the fluctuations in the CMB we observe. In Sect. 1.4, I will describe reheating, a stage right after inflation where most elementary particles were produced. I will finish by discussing gravitational wave propagation, production and detection in Sect. 1.5.

The work carried out during my PhD is described in Chaps. 2 and 3. Generally speaking, the focus of the thesis is on how to use gravitational waves as a tool to uncover new physics. I will discuss two separate topics, one related to tensor perturbations from inflation, the other to GWs produced during preheating.

In Chap. 2, I will consider cosmological perturbation theory from the point of view of loop quantum gravity, where different gravitational variables to the usual ones are used. Although classically this does not make a difference, quantum mechanical quantities are affected by this choice. I will show how this might lead to a chirality in the power spectrum of tensor perturbations from inflation. This chirality could leave a distinctive imprint on the polarization of the CMB, and we should soon be able to test whether such an effect is actually present.

In Chap. 3, I will focus on GWs produced during preheating, a non-equilibrium stage after inflation where the inflaton decays to other fields. Using a model where the inflaton is coupled to a light scalar, numerical simulations I carried out show that the GW background from this time should be anisotropic on large scales today, with relative fluctuations of order 1 %. The characteristics of this anisotropy strongly depend on the inflaton potential and its coupling to other fields, providing a novel way of constraining inflationary models.

Throughout this thesis, I will use natural units where $c = k_B = \hbar = 1$. The Planck mass will be denoted by $M_{\rm Pl} = 1/\sqrt{G} = 1.22 \times 10^{19}\,{\rm GeV/c^2}$, and the reduced Planck mass by $m_{\rm Pl} = 1/\sqrt{8\pi G} = 2.44 \times 10^{18}\,{\rm GeV/c^2}$.

1.1 The Homogeneous Universe

In this section I want to give the necessary mathematical background to describe the universe on large scales, where it looks homogeneous and isotropic, and where its expansion depends on the total matter content. I will start by introducing key aspects of general gelativity in Sect. 1.1.1 and then discuss its application to Cosmology in 1.1.2, highlighting the importance of the Friedmann equations. An accessible introduction to general relativity which also covers its use in Cosmology is Sean Carroll's *Spacetime and Geometry: An Introduction to General Relativity* [21].

1.1.1 General Relativity and Einstein's Equation

In GR, the force of gravity is a consequence of the curvature of spacetime. While in Newtonian physics we can only describe how gravity affects the motion of matter, in general relativity matter also dictates the geometry of spacetime itself [21]. This mutual relationship is described by Einstein's equation,

$$G_{\mu\nu} \equiv R_{\mu\nu} - \frac{1}{2} g_{\mu\nu} R = 8\pi G T_{\mu\nu} , \tag{1.1}$$

where the energy-momentum tensor $T_{\mu\nu}$ on the RHS describes the matter content, and the LHS the geometry of spacetime (and therefore gravity) through the Ricci tensor $R_{\mu\nu}$, which is a function of the metric tensor $g_{\mu\nu}$ and its derivatives [21].

General relativity has a rich underlying mathematical structure belonging to the field of differential geometry. A very good and thorough treatment of GR in this manner can be found in e.g. [19]. Luckily, in the context of Cosmology, we only need a basic knowledge of differential geometry to carry out calculations; this includes how tensors transform under coordinate transformations and how we can use the metric to describe spacetime and its effect on test particles. I will collect these results and their application to the universe we live in this section.

In GR, tensors are multi-linear functions defined on spacetime [21]. A rank (m, n) tensor with m upper and n lower indices maps m dual vectors and n vectors to the real numbers $\mathbb{R}$. A vector field $v = v^\mu(x^\nu)\partial_\mu$ and a dual vector field (one-form) $w = w_\mu(x^\nu)dx^\mu$ are objects living on a manifold $\mathcal{M}$, where $\mu, \nu = 1, \ldots, \dim(\mathcal{M})$ and repeated indices are summed over according to the Einstein summation convention. The components of the vector and one-form fields are $v^\mu(x^\nu)$, $w_\mu(x^\nu)$, while ∂_μ, dx^μ are a particular choice of orthogonal (coordinate) basis vectors. Compare this to ordinary three dimensional Euclidean space, where any vector field $\mathbf{v}(\mathbf{x}, \mathbf{y}, \mathbf{z})$ can be written as $\mathbf{v} = a\hat{\mathbf{x}} + b\hat{\mathbf{y}} + c\hat{\mathbf{z}}$, where a, b, c are real functions and $\hat{\mathbf{x}}, \hat{\mathbf{y}}, \hat{\mathbf{z}}$ are orthogonal unit basis vectors.

We can act with vectors on dual vectors to obtain a real number and vice versa. Rank zero and $(1, 0)$ tensors therefore correspond to ordinary scalars and vectors, respectively. Tensors can be manipulated using their components $T^{\mu_1 \cdots \mu_m}{}_{\nu_1 \cdots \nu_n}(x^\alpha)$

(where indices are raised and lowered using the metric tensor $g^{\mu\nu}$ or $g_{\mu\nu}$) with respect to a basis for the tangent space of the manifold, and I will usually refer to the components simply as the tensor. A general tensor with m upper and n lower indices transforms under a change of coordinates $x^\alpha \to x^{\alpha'}$ as [21]

$$T^{\mu'_1\cdots\mu'_m}{}_{\nu'_1\cdots\nu'_n}(x^{\alpha'}) = \frac{\partial x^{\mu'_1}}{\partial x^{\mu_1}}\cdots\frac{\partial x^{\mu'_m}}{\partial x^{\mu_m}}\frac{\partial x^{\nu_1}}{\partial x^{\nu'_1}}\cdots\frac{\partial x^{\nu_n}}{\partial x^{\nu'_n}}T^{\mu_1\cdots\mu_m}{}_{\nu_1\cdots\nu_n}(x^\alpha)\,. \quad (1.2)$$

Scalars (having no indices) do not change under a change of coordinates, $\phi'(x^{\alpha'}) = \phi(x^\alpha)$.

Let us go back to the components of the Einstein equation (1.1). The metric encodes the notion of distance in spacetime [21]. In particular, the line element

$$ds^2 = g_{\mu\nu}dx^\mu dx^\nu\,, \quad (1.3)$$

measures the proper time $d\tau$ (where $d\tau^2 = -ds^2$) experienced by an object moving an infinitesimal distance dx^μ in spacetime (this is in contrast to the coordinate time dt, which depends on the coordinates used). Note that $ds^2 < 0$ for timelike separated points (ones within the lightcone, whose interior describes the causally connected region), $ds^2 > 0$ for spacelike separated points and $ds^2 = 0$ for null separated points (which are connected by a photon trajectory) [21]. Trajectories of particles through spacetime can only connect timelike or null separated points, otherwise they would have to travel at a speed faster than the speed of light, violating special relativity.

In curved space, partial derivatives alone cannot describe variations in tensorial quantities between different coordinate patches. Instead, we need to define covariant derivatives which act on vectors v^α as

$$\nabla_\mu v^\alpha = \partial_\mu v^\alpha + \Gamma^\alpha_{\mu\nu}v^\nu\,, \quad (1.4)$$

where $\Gamma^\alpha_{\mu\nu}$ is an object called the connection which ensures that the covariant derivative transforms as a tensor [21]. The connection chosen in GR is symmetric in the lower indices (torsion free) and satisfies $\nabla_\alpha g_{\mu\nu} = 0$ (metric compatible). It is called the Christoffel connection [21],

$$\Gamma^\alpha_{\mu\nu} = \frac{1}{2}g^{\alpha\beta}(\partial_\mu g_{\nu\beta} + \partial_\nu g_{\beta\mu} - \partial_\beta g_{\mu\nu})\,. \quad (1.5)$$

The connection can be used to build the Riemann curvature tensor $R^\alpha{}_{\beta\mu\nu} = \partial_\mu\Gamma^\alpha_{\nu\beta} - \partial_\nu\Gamma^\alpha_{\mu\beta} + \Gamma^\alpha_{\mu\sigma}\Gamma^\sigma_{\nu\beta} - \Gamma^\alpha_{\nu\sigma}\Gamma^\sigma_{\mu\beta}$. $R^\alpha{}_{\beta\mu\nu}$ and its contraction $R_{\mu\nu}$, the Ricci tensor, encode the curvature of spacetime.

The energy-momentum tensor $T_{\mu\nu}$ is conserved, $\nabla_\mu T^{\mu\nu} = 0$ [21]. This relation includes four separate conservation equations. According to Noether's theorem [27], any system with a symmetry will have a corresponding conserved quantity. In the case of GR, the theory is invariant under infinitesimal time and spatial translations, with the former leading to energy conservation, $\nabla_\mu T^{\mu 0} = 0$, and the latter to momentum conservation, $\nabla_\mu T^{\mu i} = 0$ [19]. These are continuity equations for the 00 and $0i$ components of $T_{\mu\nu}$, which in flat space (where $\nabla_\mu T^{\mu\nu} = \partial_\mu T^{\mu\nu} = 0$) can be identified with the energy and momentum density, respectively.

In flat space with no gravity (where the laws of special relativity hold), the appropriate metric is the Minkowski metric $\eta_{\mu\nu}$ which is diagonal with elements $(-1, 1, 1, 1)$. This form of the metric is important on scales where the expansion of the universe can be ignored and the spacetime therefore looks flat.

In Cosmology, we will often be interested in the energy-momentum tensor of a perfect fluid [20],

$$T^{\mu\nu} = (\rho + P)u^\mu u^\nu + P g^{\mu\nu}\,, \tag{1.6}$$

where $u^\mu = dx^\mu/d\tau$ is the normalized 4-velocity of the fluid, $u_\mu u^\mu = -1$. In the local rest frame where we can use the Minkowski metric, the 3-velocity vanishes, $u^i = dx^i/d\tau = 0$, and hence $u^\mu = (1, 0)$. We can plug this into Eq. (1.6) to find $T^{00} = \rho$, $T^{0i} = 0$ (as it measures the momentum density, which is zero in the rest frame) and $T^{ij} = P\delta^{ij}$, so the fluid is isotropic. Otherwise, it would also contain an anisotropic stress term Π_{ij}.

It will be useful to regard GR in the Lagrangian formulation (see e.g. [21]), where Einstein's equations can be derived by minimising the action. The action is a functional that captures the dynamics of a physical system and is given by

$$S = \int d^4x\sqrt{-g}\mathcal{L}\,, \tag{1.7}$$

where $d^4x\sqrt{-g}$ is the covariant volume element [28] (note this is just unity for the Minkowski metric, i.e. the flat space volume is the familiar $d^3x dt$). The Lagrangian density $\mathcal{L}(\Phi^i, \nabla_\mu\Phi^i)$ depends on a field $\Phi^i(x^\mu)$ and its derivatives and describes the kinetic and potential energy of the system. We can get an equation of motion for the field by minimising the action under infinitesimal changes in the field, $\delta S = 0$. In GR, the action leading to the Einstein equation (1.1) is called the Einstein-Hilbert action and is given by [21]

$$S_H = \int \sqrt{-g}R\, d^4x\,, \tag{1.8}$$

where R is the Ricci scalar. Varying (1.8) with respect to the metric tensor would actually only give you Eq. (1.1) with RHS $= 0$, as we have not yet included any sources. To find the general expression, consider the Lagrangian [20]

$$\mathcal{L} = \frac{1}{2}m_{\rm Pl}^2 R + \mathcal{L}_{\rm mat}\,, \tag{1.9}$$

where $\mathcal{L}_{\rm mat}$ is due to matter fields (where matter here refers to anything that is not gravity). The energy-momentum tensor can be derived by minimising the action due to Eq. (1.9) with respect to $g_{\mu\nu}$, and we can then identify [20]

$$T_{\mu\nu} = -2\frac{\partial \mathcal{L}_{\text{mat}}}{\partial g_{\mu\nu}} + g_{\mu\nu}\mathcal{L}_{\text{mat}} \, . \tag{1.10}$$

1.1.2 The Friedmann Equations

The universe on large scales is homogeneous and isotropic, so the metric needs to reflect these properties. The spacetime satisfying these properties is described by the Friedmann-Robertson-Walker metric, FRW for short, which for general spatial 3-curvature k is given by (in spherical polar coordinates) [1]

$$ds^2 = -dt^2 + a^2(t)\left[\frac{dr^2}{1-kr^2} + r^2 d\Omega^2\right] , \tag{1.11}$$

where $d\Omega^2 = d\theta^2 + \sin^2\theta d\phi$ is the angular volume element.

The most recent CMB experiments have shown [14, 15] that we live in a flat universe with $k = 0$, with dark energy making up around 68 % of the energy density budget (which could correspond to a positive cosmological constant Λ or a dynamic field with negative pressure [25]), and about 27 % of dark matter and 5 % ordinary matter. Therefore, we are mainly interested in the flat space version of the FRW metric, which is usually expressed in Cartesian coordinates:

$$ds^2 = -dt^2 + a^2(t)[dx^2 + dy^2 + dz^2] \, . \tag{1.12}$$

The coordinates x, y, z are comoving, meaning they do not change with the expansion. Physical distances are related to comoving ones as $\Delta r(t) = a(t)\Delta x$.

It will sometimes be useful to use conformal time η instead of ordinary coordinate time, which is defined by $d\eta = dt/a$. This can be integrated to find the value of η at a time t':

$$\eta(t') = \int_0^{t'} \frac{dt}{a(t)} \, . \tag{1.13}$$

We can then rewrite the FRW metric (1.12) as

$$ds^2 = a^2(\eta)\eta_{\mu\nu}dx^\mu dx^\nu \, . \tag{1.14}$$

We can assume that the energy-momentum tensor of the universe is decribed by a perfect fluid, Eq. (1.6). As we live in an isotropic universe, the fluid's rest frame should coincide with the comoving coordinates in the FRW metric (1.12). From the Einstein equations we obtain two equations describing the evolution of the scale factor depending on the energy content of the universe, the Friedmann equations [20]:

$$H^2 = \left(\frac{\dot{a}}{a}\right)^2 = \frac{\rho}{3m_{\rm Pl}^2}\,, \tag{1.15}$$

$$\dot{H} + H^2 = \frac{\ddot{a}}{a} = -\frac{\rho + 3P}{6m_{\rm Pl}^2}\,, \tag{1.16}$$

where ρ contains the density of all species in the universe (like matter, radiation or a cosmological constant term $\rho_\Lambda = m_{\rm Pl}^2 \Lambda$ [20]), P is their pressure and the Hubble rate $H \equiv \frac{\dot{a}}{a}$ is an important physical length (time) scale in Cosmology.

The Friedmann equations can be combined into the continuity equation

$$\dot{\rho} = -3H(\rho + P)\,, \tag{1.17}$$

which corresponds to energy conservation for adiabatic expansion, which is valid in an isotropic universe [20].

Energy density and pressure can be related by an equation of state $P = w\rho$, which is a constant for ordinary species (but not for a scalar field) [29]. Specifically, $w = 0$ for matter (no pressure), $w = 1/3$ for radiation and $w = -1$ for a cosmological constant [20]. Using the continuity equation, this implies $\rho_M \propto a^{-3}$ for the energy density of matter (which dilutes as the volume of space grows), $\rho_R \propto a^{-4}$ for radiation (where the extra factor of a can be undestood as a redshift), and the energy density of Λ is constant. This means the evolution of the universe will consist of a series of epochs as the different powers of the scale factor compete in the Friedmann equations: first radiation domination, followed by matter domination and eventually dark energy (Λ) domination when the matter has been sufficiently diluted by the expansion.

Using the relation $\rho(a)$, the Friedmann equation (1.15) can be integrated and we obtain the evolution of the scale factor during the different epochs: $a \propto t^{2/3}$ for matter domination, $a \propto t^{1/2}$ for radiation domination and $a \propto e^{Ht}$ for a universe dominated by a cosmological constant Λ.

1.2 Inflation

This section is dedicated to introducing the theory of inflation, starting with the initial motivation for an accelerated stage of expansion to solve problems in the Big Bang model of Cosmology in 1.2.1. I will then describe single-field slow-roll inflation as the easiest possible implementation of the theory, Sect. 1.2.2. Having explored the homogeneous inflaton field, I will outline in Sect. 1.2.3 how to proceed when fluctuations are included and define the power spectra that encode the statistical properties of the field. Before explaining how inflation can act as a seed for all structure through the stretching of quantum fluctuations to cosmological scales (Sect. 1.2.5), I will need to give some details on the canonical quantization procedure and its application to curved spacetime in 1.2.4.

1.2.1 The Horizon Problem

The Big Bang model of the universe is very successful at explaining how we come to live in in expanding universe [25]. However, three separate observations show that there is something missing in our understanding of the early universe [20]. The most important one to understand conceptually is the horizon problem, so I will explain its significance and how it can be resolved using inflation. At the end, I will briefly mention the related flatness and monopole problems for completeness. For much more detailed information about the theory of inflation, see [20] or [25].

A very important concept in Cosmology is the particle horizon, the distance travelled by a photon between $t = 0$ to t' [25]. As for a photon $ds^2 = 0$, using Eq. (1.11), we can express this distance in comoving units as

$$\eta = \int_0^{t'} \frac{dt}{a(t)} \,. \tag{1.18}$$

As it has the same form as conformal time, we use the same symbol η to denote it. Clearly this quantity must always increase (as $a > 0$), and points in space separated by distances larger than the comoving horizon have non-intersecting past lightcones, i.e. no signal could have ever been transmitted between them: the points are "causally disconnected" [25] (although particles located at such points might come into causal contact in the future as the comoving horizon grows, when enough time has passed for photons from one particle to reach the other).

With this in mind, the uniformity of the CMB presents a mystery. Assume that the universe has always been matter dominated, such that $a = (t/t_0)^{2/3}$ and $H = \frac{2}{3}t^{-1}$, where the subscript zero refers to quantities today and we normalise $a_0 = 1$. We can then derive the comoving distance a photon has travelled at scale factor a_* using Eq. (1.18):

$$\eta = \int_0^{t_*} t_0^{2/3} t^{-2/3} dt = 3 t_0^{2/3} t_*^{1/3} = 2 H_0^{-1} \sqrt{a_*} \,. \tag{1.19}$$

Therefore, the comoving particle horizon at the time of recombination (when neutral hydrogen starts to form and the universe ceases to be opaque), when $a \approx 1100$, is a factor of $\sqrt{1100}$ smaller than it is today. Indeed, you can show that points with an angular separation of more than about 1° on the sky today were causally disconnected at the time of recombination [20]. However, we observe the CMB to be close to uniform on all scales. It seems very surprising that photons, free-streaming since recombination, which come from regions that were separated by distances larger than the particle horizon should just happen to be at nearly the same temperature, although no physical process could have led them to equilibrate [20].

This problem can be resolved if, before the period of radiation domination, the expansion occurred in an "unusual" manner, in which the increase of the particle horizon does not imply that the size of causally connected regions grows. This can be achieved by a period of accelerated expansion where points in space move away from each other so fast that a photon cannot traverse the distance between them. Regions that were causally connected early on could therefore move out of "causal contact", so a signal transmitted from one point will not reach another point in the region again until some time far in the future, when it has had enough time to traverse the distance between them that has grown exponentially due to the expansion.

Thus, if initially causally connected regions have been stretched to a size larger than the surface of last scattering (the surface in spacetime the CMB photons we observe originated from), the uniformity of the CMB temperature is not surprising, as the whole observable universe could have originated from a small homogeneous patch [20].

To make this more mathematically rigorous, let me define the comoving Hubble radius $(aH)^{-1}$. This is a very important length scale in Cosmology, and is often referred to simply as the (comoving) horizon. Note that unlike the particle horizon, it is not an actual horizon, and the terminology can be confusing. We can rewrite Eq. (1.18) in terms of this quantity,

$$\eta = \int_0^{a'} d(\ln a) \frac{1}{aH} \,. \tag{1.20}$$

To understand the physical significance of the comoving Hubble radius, consider a small amount of expansion for which the comoving particle horizon grows by an amount $\Delta\eta = N(aH)^{-1}$, where $N = \Delta \ln(a)$ is the number of e-folds of expansion (which counts the factors of e the scale factor has grown by). The Hubble radius then corresponds to the distance travelled by a photon while the universe expands by N e-folds.

During matter and radiation domination, the comoving Hubble radius grows monotonically, and is actually proportional to η [21]. However, if there is a stage where $(aH)^{-1}$ shrinks, so that photons traverse smaller and smaller distances during the same amount of expansion, the particle horizon can still grow (with the main contribution to the integral coming from early times) but the size of the region in causal contact at the end of this phase is much smaller than it was initially.

As the particle horizon becomes very large early on if the comoving Hubble radius shrinks, we have solved the horizon problem: At the time of recombination, the particle horizon is much larger than the distance travelled by photons since then. Therefore, when we observe the CMB, we see photons from a region whose spatial extent was within the physical horizon, which means that they could have all been at nearly the same temperature.

The condition for the comoving radius to shrink is equivalent to accelerated expansion,

$$\frac{d}{dt}(aH)^{-1} = \frac{d}{dt}\frac{1}{\dot{a}} < 0 \quad \Leftrightarrow \quad \ddot{a} > 0 \,. \tag{1.21}$$

Inflation describes this period of a shrinking comoving Hubble radius. Before describing what conditions need to be satisfied to lead to this behaviour, let me mention the other two problems inflation solves: the flatness and the monopole problem. For a curved FRW metric (1.11) with $k \neq 0$, the Friedmann equation (1.15) has an additional term proportional to the curvature and can be rewritten as [20]

$$\Omega(a) - 1 = \frac{k}{a^2 H^2}, \tag{1.22}$$

where $\Omega(a) = \rho(a)/\rho_{\text{crit}}(a)$ and $\rho_{\text{crit}}(a)$ is the density for a flat FRW universe. For perfect flatness $k = 0$, we need $\Omega(a) = 1$. However, without inflation, any small deviation from flatness will be amplified with time as $(aH)^{-1}$ grows. As we observe near flatness today, this means that Ω must have been extremely fine tuned. Inflation circumvents this as it drives $\Omega \to 1$ in Eq. (1.22) while the comoving horizon shrinks, thus solving the flatness problem.

The monopole problem was actually one of the initial motivations for inflation [9]. Grand unified theories predict the existence of unwanted relics such as magnetic monopoles, which we do not currently observe. However, the fast expansion during inflation can vastly reduce the density of these relics [20].

1.2.2 Single Field Slow-Roll Inflation

How can we satisfy the conditions required for inflation? From the second Friedmann equation (1.16) we see that accelerated expansion implies $\rho + 3P < 0$, i.e. we need a material with negative pressure driving the expansion. Sato, Kazanas and Guth [7–9] first realised that a scalar field with a specific form of the potential could satisfy this condition. The Lagrangian for a scalar field $\phi(x^\mu)$ is given by [20]

$$\mathcal{L}_\phi = -\frac{1}{2}\partial_\mu\phi\partial^\mu\phi - V(\phi). \tag{1.23}$$

The stress-energy tensor for a scalar field can then be deduced using Eq. (1.10):

$$T_{\mu\nu} = \partial_\mu\phi\partial_\nu\phi - g_{\mu\nu}\left(\frac{1}{2}\partial_\alpha\phi\partial^\alpha\phi + V(\phi)\right). \tag{1.24}$$

We can derive an equation of motion for ϕ from the Euler-Lagrange equations, i.e. from varying the action by $\delta\phi$. This leads to a wave equation [26]

$$\ddot{\phi} + 3H\dot{\phi} - \nabla^2\phi + \frac{dV}{d\phi} = 0. \tag{1.25}$$

For now, we will be concerned with a homogeneous field for which $\nabla_i \phi = 0$, such that it only depends on time, $\phi = \phi(t)$. We can then locally go to a frame with $g_{\mu\nu} = \eta_{\mu\nu}$, where the momentum density vanishes [20]. The 00 and ij components of the stress energy tensor (1.24) can then be simply identified with the rest energy density and pressure as in Sect. 1.1.1. This gives

$$\rho_\phi = \frac{1}{2}\dot{\phi}^2 + V(\phi)\,, \tag{1.26}$$

$$P_\phi = \frac{1}{2}\dot{\phi}^2 - V(\phi)\,. \tag{1.27}$$

From Eqs. (1.26) to (1.27) we see that we can satisfy the condition of negative pressure if the potential energy dominates the kinetic energy, i.e. $\dot{\phi}^2 \ll V(\phi)$, for which we obtain $w_\phi = \frac{P_\phi}{\rho_\phi} \simeq -1$. In the original models [7–9], this was achieved by trapping the inflaton field in a false minimum, with a large potential energy. While it is trapped, the universe inflates, until the inflaton spontaneously tunnels to the genuine global vacuum. However, it was found that this process could not happen in different regions of the universe fast enough to be in agreement with observations [25].

The "new" slow-roll inflation scenario requires the scalar field to slowly roll down a flat potential [30–32]. We need to satisfy two conditions: $\dot{\phi}^2 \ll V(\phi)$, to obtain accelerated expansion, and $|\ddot{\phi}| \ll |3H\dot{\phi}|, |V_{,\phi}|$, which ensures inflation lasts long enough [29] by preventing the inflaton from simply rolling down to the bottom of its potential. These conditions can be quantified by two slow-roll parameters [33],

$$\epsilon = \frac{m_{\rm Pl}^2}{2}\left(\frac{V'}{V}\right)^2\,, \quad \eta = m_{\rm Pl}^2 \frac{V''}{V}\,, \tag{1.28}$$

where primes denote differentiation with respect to ϕ and the slow-roll regime corresponds to ϵ, $|\eta| \ll 1$.

Note that during slow-roll inflation, the Friedmann equation (1.15) reduces to

$$H^2 \simeq \frac{1}{3m_{\rm Pl}^2} V(\phi)\,, \tag{1.29}$$

which implies $H \approx$ const. when $\epsilon \ll 1$. This can be easily solved to give $a(t) \sim e^{Ht}$ which is the solution for de Sitter spacetime [34], which describes a universe dominated by a cosmological constant [1]. Obviously, as the exponential expansion has to finish eventually, we cannot be in a pure de Sitter universe. Inflation ends when the slow-roll conditions are not satisfied anymore, which happens when the inflaton approaches the minimum of its potential. Any reasonable model of inflation needs to provide a mechanism for the accelerated expansion to stop, which is referred to as the graceful exit problem [26].

Let me give an example of what the slow-roll conditions imply for a specific inflaton potential. The simple quadratic potential $V = \frac{1}{2}m^2\phi^2$ is still just about viable according to the most recent Planck data [16]. From the first slow-roll condition in (1.28), we obtain $\phi > m_{\rm Pl}$, i.e. super-Planckian field values are needed for inflation to occur. Furthermore, the second slow-roll parameter η implies $m_{\rm Pl}^2 m^2 \ll V$. As the potential dominates the energy density, using Eq. (1.29) we can express the potential in terms of the Hubble rate which yields $m^2 \ll H^2$. We see that the inflaton has to be light compared to the Hubble rate for the slow-roll condition to be satisfied.

A useful quantity to consider is the number of e-foldings $N(t)$, which measures the amount of expansion during inflation. It is defined as

$$N(t) = \ln \frac{a(t_{\rm end})}{a(t)} = \int_t^{t_{\rm end}} H dt \approx \frac{1}{m_{\rm Pl}^2} \int_\phi^{\phi_{\rm end}} \frac{V}{V'} d\phi\,, \tag{1.30}$$

where the approximate equality holds during slow roll. This quantity is zero at the end of inflation and $N(t)$ therefore corresponds to the number of e-folds before the end. To agree with observations, we need the total number of e-folds $N^{\rm tot} \gtrsim 60$ in simple slow-roll models. This ensures that the largest currently observable scales were inside the Hubble horizon during inflation: The comoving Hubble scale $(aH)^{-1}$ today is e^{60} times larger than at the end of inflation [20], and comoving scales of order of the current Hubble scale must have been subhorizon during inflation, providing a lower bound for the number of e-folds [remember that during inflation, $H =$ const., and therefore it does not appear in Eq. (1.30)].

1.2.2.1 General Models of Inflation

Although I will only consider simple, monomial potentials, let me very briefly give an overview of the general classes of inflationary models that cosmologists study.

The simple single field models I just described were introduced by Linde [30] and are referred to as chaotic inflation. In these scenarios, the initial conditions in the universe are chaotic; in some regions the inflaton is displaced sufficiently far from its minimum such that it can satisfy the slow-roll conditions and lead to accelerated expansion [26]. The potential in this case can be either given by a monomial, $V(\phi) \propto \phi^p$ (where p is even due to symmetry and $p \leq 4$ to ensure renormalizability), or by an exponential, $V(\phi) \propto \exp\left(\sqrt{\frac{2}{p m_{\rm Pl}^2}}\phi\right)$ (where $p > 1$, such that the slow-roll parameters are $\epsilon = \eta/2 = 1/p$). These models require a minimum duration of the inflationary phase of $N = 60$ e-folds and a super Planckian initial field value, $\phi_{\rm start} \gg m_{\rm Pl}$ [26]. Hence, they are referred to as large-field models.

Large field models suffer from the problem that at these scales there might be correction terms to the inflaton potential which could prevent inflation from happening [35]. Therefore, models for which the inflationary phase happened at lower energy scales were introduced. These include hybrid inflation, where two scalar fields are present [36], or models inspired by supersymmetric theories [37].

Except for providing a way to drive the expansion and to end it, all models of inflation need to be consistent with the constraints from CMB measurements. Large field models typically lead to the production of an observable amount of primordial gravitational waves, see Sect. 1.2.5. This is in contrast to the small field models motivated by high energy physics [20]. Although we do not currently understand physics at super Planckian scales, the BICEP2 results [17] suggest that (modifications of) chaotic inflation models could indeed be viable.

1.2.3 Beyond the Homogeneous Field Evolution

Although inflation was initially introduced as a way of solving the Big Bang puzzles highlighted in Sect. 1.2.1, its main power and appeal lies in the fact that it can explain the origin of the primordial fluctuations that were the seed of all structure, and which we can still observe in the CMB [20]. To see this, we need to go beyond the homogeneous description and consider inhomogeneous fluctuations around the background, i.e. expand the inflaton field as:

$$\phi(\mathbf{x}, t) = \phi(t) + \delta\phi(\mathbf{x}, t)\,. \tag{1.31}$$

The homogenous part $\phi(t)$ (which is averaged over fluctuations and therefore independent of position) is responsible for the de Sitter-like expansion of the universe, whereas the fluctuations $\delta\phi(\mathbf{x}, t)$ are coupled to the fluctuations in the metric around the Friedmannian background.

In cosmological perturbation theory (see appendix A.1 for more details), we expand the metric and energy-momentum tensor to first order in perturbations. Using the SVT decomposition (appendix A.1), which describes how a general perturbation in the metric can be split into scalars, vectors and tensors that all transform differently under rotations, we find that the scalar perturbations are directly coupled to the perturbations $\delta\phi$. Vector perturbations are not produced by inflation and decay [29], while tensor perturbations correspond to GWs and are not coupled to the inflaton [20]. Instead, they need a source with non-zero quadrupole moment (see Sect. 1.5.2) which is related to a non-zero anisotropic stress in the energy-momentum tensor, not present for a scalar field [25]. Similarly, no perfect fluid can act as source for GWs [1].

1.2.3.1 Power Spectra

In the next two sections, I will explain how quantum vacuum fluctuations during inflation can become "classical" once they leave the horizon and lead to a scale-invariant spectrum of fluctuations. A power spectrum describes the amplitude of different Fourier modes k of a field ϕ and is defined as an ensemble average of the fluctuations [29]:

$$\langle \phi_{\mathbf{k}} \phi_{\mathbf{k}'} \rangle = (2\pi)^3 \delta(\mathbf{k} + \mathbf{k}') P_\phi(k) \,. \tag{1.32}$$

The power spectrum is the Fourier transform of the real space correlation function, $\langle \phi(\mathbf{x}) \phi(\mathbf{y}) \rangle$ [29]. The mean fluctuations in all fields are zero on average, i.e. $\langle \phi_{\mathbf{k}} \rangle = 0$, as there should be equally many regions with higher or lower amplitude. The power spectrum, on the other hand, gives you a statistical measure of the fluctuations as it estimates the typical deviation from the mean you would expect for each mode. A real field distribution is a realization of the statistical ensemble, and therefore drawn from the probability distribution in (1.32) [20].

We can also define a dimensionless power spectrum (denoted by a curly $\mathcal{P}$)

$$\mathcal{P}_\phi(k) = \frac{k^3}{2\pi^2} P_\phi(k) \,. \tag{1.33}$$

This is directly related to the real space variance σ_ϕ^2 of the field and describes the power per logarithmic k interval [29]:

$$\sigma_\phi^2 \equiv \langle \phi(\mathbf{x})^2 \rangle = \int_0^\infty \mathcal{P}_\phi(k) d \ln k \,. \tag{1.34}$$

1.2.4 Canonical Quantization

In Sect. 1.2.5, I want to study the evolution of quantum fluctuations during inflation. To do this, we need to use the standard approach of canonical quantization introduced by Dirac [38]. It proceeds by promoting fields to operators that satisfy commutation relations (defining an algebra) which makes it possible to define particle states in terms of eigenstates of the Hamiltonian of the system.

1.2.4.1 Flat Background

Let me outline the quantization procedure for a scalar field $\phi(\mathbf{x}, t)$ on a flat background first, before generalising it to a curved background (as needed in Cosmology) in the next section. I will follow the very clear treatment in [28], working in the Heisenberg picture where operators are time-dependent and states constant.

The Lagrangian $\mathcal{L}_\phi$ of a scalar field is given by Eq. (1.23). Let us assume from now on that we are dealing with a free field, which does not interact with other fields or itself. In this case, the potential is $V(\phi) = \frac{1}{2} m^2 \phi^2$ [28]. The action (1.7) for a scalar field on a flat background is then

$$S = \frac{1}{2} \int d^3x dt \left[\dot{\phi}^2 - (\nabla \phi)^2 - m^2 \phi^2 \right] \,. \tag{1.35}$$

Just like in classical mechanics, we can use the Hamiltonian instead of the Lagrangian to describe the dynamics of the system. They are related by

$$H = \int d^3x \left[\pi\dot{\phi} - \mathcal{L}\right], \tag{1.36}$$

where $\pi(\mathbf{x}, t) \equiv \frac{\partial \mathcal{L}}{\partial \dot{\phi}(\mathbf{x},t)}$ is the conjugate momentum density. Using Eq. (1.23), $\pi = \dot{\phi}$ and the Hamiltonian becomes

$$H = \frac{1}{2}\int d^3x \left[\pi^2 + (\nabla\phi)^2 + m^2\phi^2\right]. \tag{1.37}$$

The equation of motion, derived by minimising the action or from Hamilton's equations, is just

$$\ddot{\phi} - \nabla^2\phi + m^2\phi = 0. \tag{1.38}$$

We can expand the field in terms of Fourier modes,

$$\phi(\mathbf{x}, t) = \int \frac{d^3\mathbf{k}}{(2\pi)^3} e^{i\mathbf{k}\cdot\mathbf{x}} \phi_{\mathbf{k}}(t), \tag{1.39}$$

and similarly for $\pi(\mathbf{x}, t)$. Substituting the Fourier space expansion into the equation of motion (1.38), we find that the field satisfies the equation of a harmonic oscillator,

$$\ddot{\phi}_{\mathbf{k}} + \omega_k^2 \phi_{\mathbf{k}} = 0, \tag{1.40}$$

with frequency $\omega_k^2 \equiv k^2 + m^2$. The reason it is useful to treat the problem in Fourier space is that the different oscillators $\mathbf{k}$ decouple and are therefore independent of one another [28].

To arrive at the quantum theory, we follow the same approach one would learn in a quantum mechanics course when dealing with position and momentum operators (see e.g. [39] for an introduction to quantum mechanics in this manner): We promote the field and its conjugate momentum to operators $\hat{\phi}(\mathbf{x}, t)$ and $\hat{\pi}(\mathbf{x}, t)$ that need to satisfy commutation relations

$$\left[\hat{\phi}(\mathbf{x}, t), \hat{\pi}(\mathbf{y}, t)\right] = i\delta(\mathbf{x} - \mathbf{y}), \tag{1.41}$$

with all other commutators zero, and where the commutator is defined as $[A, B] = AB - BA$. Note that as the field ϕ is real, in Fourier space the operators need to satisfy $\hat{\phi}_{\mathbf{k}}^{\dagger} = \hat{\phi}_{-\mathbf{k}}$. As the modes $\hat{\phi}_{\mathbf{k}}$ behave like a harmonic oscillator, it is instructive to define creation and annihilation operators $a_{\mathbf{k}}^{\dagger}$ and $a_{\mathbf{k}}$, where

$$a_{\mathbf{k}} = \sqrt{\frac{\omega_k}{2}}\hat{\phi}_{\mathbf{k}}(0) + i\sqrt{\frac{1}{2\omega_k}}\hat{\pi}_{\mathbf{k}}(0). \tag{1.42}$$

Plugging the Fourier expansion (1.39) into (1.41), we find that they obey commutation relations

$$\left[a_{\mathbf{k}}, a^{\dagger}_{\mathbf{k}'}\right] = (2\pi)^3\delta(\mathbf{k} - \mathbf{k}') \,. \tag{1.43}$$

We can now perform the standard Fock space quantization [38], where the vacuum $|0\rangle$ is defined as the state annihilated by $a_{\mathbf{k}}$ ($a_{\mathbf{k}}|0\rangle = 0$), and n-particle states are defined by repeated application of the creation operator to the vacuum (where each $a^{\dagger}_{\mathbf{k}}$ creates a particle with momentum $\mathbf{k}$). The basis of the Fock space are eigenstates of the number operator $N_{\mathbf{k}} = a^{\dagger}_{\mathbf{k}}a_{\mathbf{k}}$, whose eigenvalues count the number of particles with momentum $\mathbf{k}$. The Hamiltonian (1.37), evaluated at $t = 0$ (as it does not explicitly depend on time and is therefore conserved), can be expressed in terms of the number operator as (ignoring an infinite "vacuum" energy contribution which is not important in quantum field theory [28])

$$\hat{H} = \int d^3\mathbf{k}\omega_k N_{\mathbf{k}} \,, \tag{1.44}$$

which clearly shows that the vacuum $|0\rangle$ is the state with the lowest possible energy.

We could also define time dependent creation and annihilation operators. They are similar to the expressions in Eq. (1.42) but are evaluated at a general t. The time dependent and independent (for $t = 0$) operators are related by $a^{\dagger}_{\mathbf{k}}(t) = a^{\dagger}_{\mathbf{k}}e^{i\omega_k t}$, $a_{\mathbf{k}}(t) = a_{\mathbf{k}}e^{-i\omega_k t}$.

In light of this, we can consider quantization from a different but equivalent viewpoint: Instead of expressing the annihilation and creation operators as linear combinations of the field and its conjugate, we simply perform a mode expansion of the field [28] as

$$\hat{\phi}_{\mathbf{k}}(t) = \left[v^*_k(t)a_{\mathbf{k}} + v_k(t)a^{\dagger}_{-\mathbf{k}}\right] \,. \tag{1.45}$$

The complex mode functions carry all the time dependence and satisfy the equation of motion

$$\ddot{v}_k + \omega_k^2 v_k = 0 \,, \tag{1.46}$$

from Eq. (1.40). A general solution to this equation is given by

$$v_k(t) = A\left(\alpha_k e^{i\omega_k t} + \beta_k e^{-i\omega_k t}\right) \,, \tag{1.47}$$

where A is a normalization factor. Like the frequency ω_k, the mode functions only depend on the magnitude $|\mathbf{k}|$, and the directional dependence is contained in the factors $a_{\mathbf{k}}, a^{\dagger}_{-\mathbf{k}}$, which can now be simply regarded as field amplitudes. Plugging the mode expansion (1.45) into (1.39) we obtain (after changing variables from $-\mathbf{k} \to \mathbf{k}$ in the second term)

$$\hat{\phi}(\mathbf{x}, t) = \int \frac{d^3\mathbf{k}}{(2\pi)^3}\left[v^*_k(t)a_{\mathbf{k}}e^{i\mathbf{k}\cdot\mathbf{x}} + v_k(t)a^{\dagger}_{\mathbf{k}}e^{-i\mathbf{k}\cdot\mathbf{x}}\right] \,. \tag{1.48}$$

We can now postulate the CRs (1.43) for the amplitudes $a_{\mathbf{k}}, a_{\mathbf{k}}^{\dagger}$. To achieve consistency with the CRs for the field and its conjugate, Eq. (1.41), the mode functions $v_k(t)$ need to satisfy the normalization condition [29]

$$\dot{v}_{\mathbf{k}}(t)v_{\mathbf{k}}^{*}(t) - v_{\mathbf{k}}(t)\dot{v}_{\mathbf{k}}^{*}(t) = i\,. \tag{1.49}$$

Equation (1.46) is a second order differential equation for the complex mode functions, so the normalization condition does not suffice to specify them completely. To determine $v_k(t)$ uniquely, we can demand that the vacuum, defined by $a_{\mathbf{k}}|0\rangle = 0$, is an eigenstate of the Hamiltonian with minimal energy. This fixes the mode functions to be [29]

$$v_k(t) = \frac{1}{\sqrt{2\omega_k}}e^{-i\omega_k t}\,. \tag{1.50}$$

Having found an expression for the mode functions, we can define vacuum fluctuations of the field $\phi(\mathbf{x}, t)$, by studying the expectation value $\langle\phi_{\mathbf{k}}\phi_{\mathbf{k}'}\rangle$ of Fourier modes in the ground state. Using equation (1.45) and noting that $a_{\mathbf{k}}|0\rangle = 0$, $\langle 0|a_{\mathbf{k}}^{\dagger} = 0$, we obtain

$$\langle\phi_{\mathbf{k}}\phi_{\mathbf{k}'}\rangle = |v_k(t)|^2\langle 0|a_{\mathbf{k}}a_{-\mathbf{k}}^{\dagger}|0\rangle\,. \tag{1.51}$$

Using Eq. (1.43), this gives

$$\langle\phi_{\mathbf{k}}\phi_{\mathbf{k}'}\rangle = (2\pi)^3\delta(\mathbf{k}+\mathbf{k}')|v_k(t)|^2\,. \tag{1.52}$$

1.2.4.2 Expanding Background

The approach involving mode functions is also used to quantize fields on a curved background. Let us consider a flat ($k = 0$) Friedmann universe where the metric is expressed using conformal time, Eq. (1.14). In these coordinates, the metric is related to flat Minkowski space by a conformal transformation [28]. This already tells us that the problem in a homogeneous isotropic curved background will be similar to the flat situation we considered above, however there will be some important differences.

Noting that $\sqrt{-g} = a^4$ for this metric, and that indices are now raised and lowered with $g_{\mu\nu} = a^2\eta_{\mu\nu}$, we see from simple substitution that the action (1.7) becomes

$$S = \frac{1}{2}\int d^3x d\eta a^2\left[\phi'^2 - (\nabla\phi)^2 - m^2a^2\phi^2\right], \tag{1.53}$$

where a prime denotes differentiation with respect to to conformal time. To make this look more like Eq. (1.35), define an auxiliary field $\chi \equiv a\phi$. Then, Eq. (1.53) can be written as [28]

$$S = \frac{1}{2}\int d^3x d\eta\left[\chi'^2 - (\nabla\chi)^2 - \left(m^2a^2 - \frac{a''}{a}\right)\chi^2\right]. \tag{1.54}$$

This looks exactly like the action for a field in flat spacetime, Eq. (1.35), except that the effective mass $m_{\text{eff}}^2(\eta) = m^2a^2 - \frac{a''}{a}$ is now time dependent. The equation of motion for the Fourier modes $\chi_{\mathbf{k}}$ derived from this action is given by [c.f. Eq. (1.40)]

$$\chi_{\mathbf{k}}'' + \omega_k^2 \chi_{\mathbf{k}} = 0\,, \tag{1.55}$$

where the frequency $\omega_k^2(\eta) = k^2 + m_{\text{eff}}^2(\eta)$ is time-dependent.

Due to the similarities in the form of the equations, we can follow the same steps to quantization as for the flat case. We can start by performing a mode expansion for $\hat{\chi}(\mathbf{k}, \eta)$ as done in Eq. (1.45), but the equation of motion for the mode functions now has the time dependent frequency $\omega_k(\eta)$.

The explicit time dependence in the action leads to complications when trying to determine the mode functions. While the same normalization condition still holds [28], the second condition, related to the choice of vacuum, does not give a well-defined answer anymore. Like the action, the Hamiltonian for the field $\hat{\chi}(\mathbf{x}, \eta)$ is now explicitly time dependent. Therefore, it cannot possess time independent eigenvectors. In particular, this means that there is no uniquely defined vacuum state.

Choosing the correct mode functions will depend on the problem at hand, and there are approaches such as using the instantaneous vacuum state, defined at a specific time, or the so-called adiabatic vacuum if the frequency $\omega_k(\eta)$ varies slowly [28]. Fortunately, in the case of inflation, the background space can be approximated by de Sitter and there is a preferred choice for the mode functions, described by the Bunch-Davies vacuum [40]. We will now consider the quantization of metric perturbations during inflation.

1.2.5 Quantum Fluctuations During Inflation

Before deriving the mode functions (and therefore the vacuum fluctuations) of the metric perturbations during inflation, let me make the following observation. While in the previous section we quantized the "full" scalar field ϕ, ignoring its interaction with the background, in the case of the inflaton we need to consider the homogeneous field and its fluctuations separately. Consider equation (1.31). The background $\phi(t)$ can be regarded as behaving completely classically; it simply drives the expansion and determines the de Sitter-like background evolution. When deriving the quantum fluctuations from inflation, we need to consider the inflaton fluctuations $\delta\phi(\mathbf{x}, t)$, which are related to the fluctuations in the metric through Einstein's equation.

In linear perturbation theory, where the perturbations are small, we can ignore all terms that are second order or higher. This is the approach used in Cosmology, and hence the perturbations can be thought of living on a fixed, unperturbed spacetime, as any backreaction effects would be of higher order in the perturbation [20].

To study quantum fluctuations during inflation, we consider the metric perturbations as free fields propagating on a fixed FRW background [29]. The scalar perturbations of the metric are coupled to the inflaton perturbations $\delta\phi(\mathbf{x}, t)$, and we can pick different gauges to study them (see appendix A.1). After inflation, they will induce density perturbations in the matter distribution and can therefore be regarded as the source of all structure.

A simpler problem to consider are the tensor perturbations, as they do not couple to inflaton perturbations [25]. As we will see explicitly in Sect. 1.5.1, the transverse and traceless spatial metric perturbations h_{ij} correspond to GWs. As GWs are the main focus of this thesis, I will outline the quantization procedure in this case, deriving a power spectrum of fluctuations of tensor modes. For completeness, I will give the result for scalar modes at the end.

To quantize the tensor perturbations h_{ij}, the Einstein-Hilbert action (1.8) in a Friedmann universe needs to be expanded to second order [29]. Keeping only the second order term yields [1]

$$S^{(2)}_{EH} = \frac{m_{\rm Pl}^2}{8} \int d\eta d^3x a^2 \left[(h'_{ij})^2 - (\partial_l h_{ij})^2 \right] . \tag{1.56}$$

As the tensor field h_{ij} contains two independent polarizations $r = +, \times$ (Sect. 1.5.1), it is useful to transform to Fourier space where it can be expressed in terms of the polarization tensor $\epsilon^r_{ij}(\mathbf{k})$, which satisfies $k^i \epsilon_{ij} = 0$, $\epsilon^r_{ij}(\mathbf{k})\epsilon^{r'}_{ij}(\mathbf{k}) = 2\delta_{r,r'}$ [41]:

$$h_{ij} = \int \frac{d^3k}{(2\pi)^3} \sum_r \epsilon^r_{ij}(\mathbf{k}) h^r_{\mathbf{k}}(\eta) e^{i\mathbf{k}\cdot\mathbf{x}} . \tag{1.57}$$

If we also define $\tilde{h}^r_{\mathbf{k}} \equiv \frac{a}{2} m_{\rm Pl} h^r_{\mathbf{k}}$ and substitute expansion (1.57) into the action (1.56) we obtain [29]

$$S^{(2)}_{EH} = \sum_r \frac{1}{2} \int d\eta d^3k \left[(\tilde{h}^{r\,\prime}_{\mathbf{k}})^2 - \left(k^2 - \frac{a''}{a}\right) (\tilde{h}^r_{\mathbf{k}})^2 \right] . \tag{1.58}$$

The corresponding expression for the Hamiltonian is given by

$$H^{(2)}_{EH} = \sum_r \frac{1}{2} \int d^3k \left[(\tilde{h}^{r\,\prime}_{\mathbf{k}})^2 + \left(k^2 - \frac{a''}{a}\right) (\tilde{h}^r_{\mathbf{k}})^2 \right] . \tag{1.59}$$

The action (1.58) is the same as two copies of Eq. (1.54) in Fourier space, but with no mass term. When quantizing the tensor perturbations, it is therefore the same exercise as trying to quantize two massless scalar fields in curved spacetime. If we expand the Fourier components in terms of creation and annihilation operators as in Eq. (1.45), we find that the mode functions obey [1]

$$v''_k + \left(k^2 - \frac{a''}{a}\right) v_k = 0 . \tag{1.60}$$

This is known as the Mukhanov equation. During inflation, we are in a quasi de Sitter phase where $H = \frac{\dot{a}}{a} = \text{const}$. Changing to conformal time, this implies $a' = a^2 H$ and integrating gives

$$a(\eta) = -\frac{1}{H\eta}\,. \tag{1.61}$$

Note that during de Sitter expansion, conformal time is negative, and becomes infinite in the past when $a \to 0$. Therefore, we can write $\frac{a''}{a} = \frac{2}{\eta^2}$ and Eq. (1.60) becomes

$$v_k'' + \left(k^2 - \frac{2}{\eta^2}\right) v_k = 0\,. \tag{1.62}$$

In de Sitter space, there exists a preferred quantum state, the Bunch-Davies vacuum. It is time independent and can therefore be used to determine unique mode functions [28]. Let us construct them. In the far past, when $|k\eta| \gg 1$, the second term in Eq. (1.62) becomes negligible. At these early times, all scales were far inside the horizon and did not feel the curvature of spacetime, so we obtain the mode equation for Minkowski space, Eq. (1.50) with $\omega_k = k$. The Bunch-Davies vacuum therefore corresponds to the minimal excitation state in the far past [28]. This condition and the normalization (1.49) are sufficient to determine the mode functions uniquely. The general solution of equation (1.62) gives the Bunch-Davies mode functions,

$$v_k(t) = \frac{e^{-ik\eta}}{\sqrt{2k}}\left(1 - \frac{i}{k\eta}\right)\,. \tag{1.63}$$

To determine the power spectrum, remember from equation (1.52) that we simply need to calculate the modulus squared of the mode functions which in this case is given by

$$|v_k|^2 = \frac{1}{2k^3\eta^2}(k^2\eta^2 + 1)\,. \tag{1.64}$$

Equation (1.64) is defined for $\tilde{h}^r_{\mathbf{k}}$, however we are interested in the power spectrum of the physical field $h^r_{\mathbf{k}} = \frac{2}{am_{\rm Pl}}\tilde{h}^r_{\mathbf{k}}$:

$$\langle \hat{h}_{\mathbf{k}}(\eta)\hat{h}_{\mathbf{k}'}(\eta)\rangle = (2\pi)^3\delta(\mathbf{k}+\mathbf{k}')\frac{4|v_k(\eta)|^2}{a^2 m_{\rm Pl}^2} \tag{1.65}$$

$$= (2\pi)^3\delta(\mathbf{k}+\mathbf{k}')\frac{2H^2}{k^3 m_{\rm Pl}^2}(1 + k^2\eta^2)\,, \tag{1.66}$$

where in the second equality we used (1.61). The first line, Eq. (1.65), seems to imply that fluctuations decay with time due to the presence of the factor a^2. This is true for any modes deep inside the horizon, with $k\eta \gg 1$: the mode $\tilde{h}^r_{\mathbf{k}}$, rescaled by a, simply oscillates in a Minkowski vacuum, but the physical mode $h^r_{\mathbf{k}}$ decays due to the expansion of the universe. The beauty of the de Sitter-like expansion during inflation is that when the modes become superhorizon, $k\eta \ll 1$, the second term in (1.66) can be ignored and the spectrum of fluctuations approaches a constant (as $H \simeq$ const). In terms of the dimensionless power spectrum (1.33),

$$\mathcal{P}_h(k) = \frac{4}{m_{\text{Pl}}^2}\left(\frac{H}{2\pi}\right)^2\bigg|_{k=aH} . \tag{1.67}$$

The total tensor power spectrum is actually twice this value as we have to take into account both polarizations. Note that Eq. (1.67) needs to be evaluated at horizon crossing, $k = aH$. H is a constant in pure de Sitter space, and therefore the power spectrum would be perfectly scale-invariant, i.e. the same for any mode k exiting the horizon at different times. However, in a slow-roll inflationary model, we only have quasi de Sitter evolution, where H is not perfectly constant and therefore the spectrum is slightly redshifted. Modes that exit the horizon earlier will have a slightly larger amplitude, as H becomes smaller with time. This scale dependence is taken into account by evaluating the spectrum at horizon crossing, so Eq. (1.67) can be used to describe fluctuations from slow-roll inflation [29].

As the power spectrum is constant on scales $k\eta \ll 1$, fluctuations "freeze out" after they cross the horizon. This is related to the fact that on scales larger than the Hubble scale $(aH)^{-1}$, no causal physics should act [29]. We therefore do not have to worry about their behaviour until they re-enter the horizon at a later time, long after inflation. Moreover, the evolution during the de Sitter expansion makes it possible to stretch quantum fluctuations to very large scales. The power spectrum in Eq. (1.67) can therefore be regarded as a classical probability distribution for tensor modes [29]. Understanding the quantum-to-classical transition is the subject of the field of decoherence [42].

Although vacuum fluctuations are always present due to the uncertainty principle, usually their amplitude is vanishingly small on large scales [1]. Only because of the accelerated expansion, which leads to a shrinking comoving Hubble volume, do we arrive at a situation where these quantum fluctuations can be stretched to cosmologically relevant scales and retain their amplitude [28].

Inflation produces a nearly scale invariant spectrum of perturbations. I have shown this explicitly for the tensor modes, but it is also possible to do the same exercise for scalar perturbations. Unlike tensors, these couple to the inflaton perturbation, so we will need to consider the spectrum of a quantity that contains both scalar metric and inflaton perturbations. A gauge invariant choice is the comoving curvature perturbation [29]

$$\mathcal{R} = \Psi + \frac{H}{\dot{\phi}}\delta\phi , \tag{1.68}$$

where Ψ is the metric perturbation corresponding to the gravitational potential in the Newtonian gauge (see appendix A.1). $\mathcal{R}$ describes the spatial curvature on comoving (constant ϕ) hypersurfaces, as measured by an observer moving with the expansion of the universe. It is a useful quantity to consider as it is conserved outside the horizon, even after the end of inflation when the inflaton (and hence its perturbation) has decayed [20].

To find the power spectrum of $\mathcal{R}$ from inflation, we can choose to quantize either the metric or inflaton perturbations, depending on the gauge. The approach is similar to the case of tensor perturbations, but the action now contains a mass term that depends on the slow-roll parameter ϵ [29]. However, in the pure de Sitter limit, where $\epsilon \to 0$, we again obtain the mode equation for a massless field (1.62). Even for slow-roll inflation, we can use the Bunch-Davies mode functions: They are well defined as long as the inflaton is a light field, $m^2 < H^2$, which is satisfied by $\eta \ll 1$ [28]. Note, however, that for a heavy field with $m^2 > H^2$, there would be a mass term in equation (1.58) that can make the effective frequency positive [28], resulting in oscillatory behaviour, rather than the freeze-out of modes as described above. Hence, only light fields can acquire a scale-invariant spectrum of perturbations during inflation.

The power spectrum of fluctuations for the comoving curvature perturbation is

$$\mathcal{P}_{\mathcal{R}}(k) = \left.\frac{H^2}{(2\pi)^2}\frac{H^2}{\dot{\phi}^2}\right|_{k=aH}, \tag{1.69}$$

where again we need to evaluate the Hubble rate at horizon crossing to take its time dependence into account. Note that this can be derived in the spatially flat gauge, where $\Psi = 0$, from the power spectrum of inflaton fluctuations [29]:

$$\mathcal{P}_{\phi}(k) = \left(\frac{H}{2\pi}\right)^2. \tag{1.70}$$

Instead of $\mathcal{R}$, we could also have considered the curvature perturbation ζ on uniform density hypersurfaces (for which $\delta\rho = 0$). It is also gauge invariant and equal to $-\mathcal{R}$ during slow-roll inflation [29]:

$$-\zeta \equiv \Psi + \frac{H}{\dot{\rho}}\delta\rho. \tag{1.71}$$

The curvature perturbations are also equivalent on superhorizon scales $k \ll aH$ where $\zeta = -\mathcal{R}$, as long as there are only adiabatic density perturbations [43], so they can often be used interchangeably.

The scale dependence of the tensor and curvature power spectra can easily be quantified by introducing spectral indices n_s and n_t for the scalar and tensor modes [20]:

$$n_s - 1 \equiv \frac{d\ln\mathcal{P}_{\mathcal{R}}}{d\ln k} = 2\eta - 6\epsilon, \tag{1.72}$$

$$n_t \equiv \frac{d\ln\mathcal{P}_t}{d\ln k} = -2\epsilon. \tag{1.73}$$

The second equality shows the value of the spectral indices in terms of the slow roll parameters (1.28) when calculated in the slow-roll approximation. As they are very small, this demonstrates the near scale invariance of the spectra.

It is also useful to define the tensor to scalar ratio [29]

$$r \equiv \frac{\mathcal{P}_t}{\mathcal{P}_{\mathcal{R}}} = 16\epsilon \,. \tag{1.74}$$

In slow-roll inflation, scalars strongly dominate over tensors. The values of n_s and r are used to constrain inflationary models, with the most recent bounds by Planck giving $n_s = 0.9603 \pm 0.0073$ and $r < 0.11$ [16].

While the spectrum of scalar modes has been extensively probed through CMB temperature and polarization measurements (see next section), tensor modes from inflation have remained elusive for a long time. Very recently, the BICEP2 collaboration [17] detected B-mode polarization of potentially primordial origin, which remains to be verified by other experiments. Their analysis suggests a value of $r \approx 0.2$, however this has been obtained from only a small patch of sky over the South Pole and will probably change when a full sky analysis is available, which would also provide us with the full spectrum of tensor perturbations.

Tensor modes are a very useful tool in constraining models of inflation. From Eq. (1.67) it is clear that the spectrum of tensor perturbations is directly proportional to H^2 and therefore $\mathcal{P}_h \propto \rho/m_{\rm Pl}^4$ from the Friedmann equation (1.15). Hence, it gives you a direct measure of the energy scale of inflation and therefore the inflationary potential, which dominates the energy density [25].

1.3 From Inflation to the Cosmic Microwave Background

In this section I want to give a very brief overview of what happens to the primordial spectra after inflation. Studying this evolution in detail is a complicated field (rooted in cosmological perturbation theory) and the subject of many Cosmology textbooks, see e.g. [20, 25]. The scalar perturbations are the seed of all the structure we observe, while both scalar and tensor modes will leave an imprint on the CMB anisotropies. In particular, a specific pattern in the polarization of the CMB distribution, the B-mode, can only be sourced by tensor perturbations and could therefore provide a direct window into the study of gravitational waves from inflation.

1.3.1 Perturbations in Matter and Radiation

The power spectra for scalars and tensors derived in the previous section determine the subsequent evolution of perturbations. Any quantity of cosmological interest can be ultimately traced back to these initial conditions. After the inflaton decays (which is the subject of reheating, described in Sect. 1.4), the presence of the curvature perturbation $\mathcal{R}$ will source density fluctuations in each particle species [20]: baryonic matter, cold dark matter, neutrinos and photons.

The perturbations in the fields set up by the simplest slow-roll inflationary scenario are adiabatic, which means that their number densities are perturbed by the same factor [26], and Gaussian, so their Fourier components, like the vacuum fluctuation, have independent probability distributions [20] (and we therefore only need 2-point functions to describe them). More complicated models of inflation can lead to isocurvature perturbations and non-Gaussianity, but both of these features are subdominant according to the most recent data [15].

We typically study the fluctuations in Fourier space, as for small perturbations different **k** modes will evolve independently [1]. Fourier modes behave very differently depending on whether they are outside (frozen in) or inside the horizon (when causal physics can affect them [25]). Particularly, perturbations that re-enter the horizon (i.e. **k** modes with wavenumber $k > aH$) during radiation domination will evolve very differently to ones that enter during matter domination.

Another important epoch is the time of recombination around 380.000 years after the Big Bang, when neutral atoms can first form. Until this point, the photons and baryons were tightly coupled [20], and the photons' mean free path was very small as they constantly Compton scattered off free electrons. At the time of recombination, this mean free path increases until the photons completely decouple from the baryons and free-stream to us today, where we observe them as the CMB. The position of this event in spacetime is referred to as the surface of last scattering.

On very large scales, which were superhorizon at the time of recombination, we can directly relate the fractional temperature perturbation in the CMB to the primordial curvature perturbation $\mathcal{R}$ [20],

$$\frac{\delta T}{T} = -\frac{1}{5}\mathcal{R}. \tag{1.75}$$

However, on smaller scales fluctuations in the matter and radiation densities have evolved and will not simply be related to the primordial spectra anymore. Instead, the evolution is described using transfer functions $T(t)$ [20, 25], which relate the power spectrum of any field $g(\mathbf{x})$ at time t to the primordial spectrum as $\mathcal{P}_g(t) = T_g^2(t)\mathcal{P}_\mathcal{R}$.

The perturbations in the matter distribution are the seeds of all the structure we observe, from stars to clusters of galaxies. These gravitationally bound object can form when the density contrast $\delta\rho/\rho$ becomes large and the equations of motion become non-linear, so cosmological perturbation theory no longer holds. For photons, on the other hand, radiation pressure prevents perturbations from gravitationally collapsing [25]. The CMB fluctuations were therefore imprinted when the evolution was still linear, which makes it possible to check the validity of the inflationary paradigm by directly probing the primordial power spectra (and taking the transfer functions into account).

The perturbations in the photon distribution manifest themselves as fractional temperature perturbation in the CMB, commonly denoted by $\Theta(\mathbf{x}, \hat{p}, t) = \delta T/T$ [25].

The photon distribution is mathematically more complicated than non-relativistic matter, as it depends not only on time t and position $\mathbf{x}$, but also the direction of propagation of the photons $\hat{p}$ (so the CMB is not only inhomogeneous, but also anisotropic [25]). It makes sense to expand the photon distribution function in terms of Legendre polynomials $\mathcal{P}_l$ to take care of the $\hat{p}$ dependence [25]. Specifically, defining $\hat{k} \cdot \hat{p} = \cos\theta$, in Fourier space we get

$$\Theta_l(\mathbf{k}, t) \equiv \frac{1}{2(-i)^l} \int_{-1}^{1} d(\cos\theta)\, \mathcal{P}_l(\cos\theta)\Theta(\mathbf{k}, \cos\theta, t)\,, \tag{1.76}$$

so the full photon distribution can be expressed as an infinite series of moments l. $l = 0$ is the monopole, $l = 1$ the dipole, $l = 2$ the quadrupole etc. When we measure the CMB at our position ($\mathbf{x}_0$) today (t_0), we can only probe the directional dependence $\hat{p}$. It is then useful to expand the perturbation in terms of spherical harmonics instead. They enable us to expand the perturbation on a sphere [25], so we can describe photons arriving at our position which originated at the last scattering surface. The amplitudes of each spherical harmonic are called the multipole moments and are given by [29]

$$a_{lm} = \int d\Omega Y_{lm}^*(\hat{p})\Theta(\hat{p})\,. \tag{1.77}$$

The label l of the spherical harmonics is related to the angular size θ of the perturbation on the last scattering surface, $\theta \sim \pi/l$ [25], so larger multipoles probe smaller angular scales.

The multipole moments fully characterise the perturbation. Their mean is zero, while their variance $\langle a_{lm}a^*_{l'm'}\rangle = C_l\delta_{ll'}\delta_{mm'}$ describes the statistical properties of the field, i.e. a typical realization of each multipole moment a_{lm} will be drawn from a Gaussian centred around zero with variance C_l [25]. It does not depend on m (which takes integer values between $-l$ and l) due to the rotational invariance of the background. The variance is related to the primordial power spectrum of curvature perturbations (1.69) (which dominates over tensors),

$$C_l = 4\pi \int_0^\infty T_\Theta^2(k, l)\mathcal{P}_\mathcal{R}(k)\frac{dk}{k}\,, \tag{1.78}$$

where $T_\Theta^2(k, l)$ is the transfer function for the temperature perturbations [20]. Equation (1.78) is the angular power spectrum of temperature fluctuations.

The angular power spectrum has been explored in great detail by WMAP [14] and, with even higher angular resolution, by Planck [15], providing us with a wealth of information about the early universe. The temperature power spectrum is dominated by scalar perturbations and cannot be used to extract parameters characterising the tensor perturbations [20]. However, primordial tensor modes can be measured through the CMB polarization [25], which I will discuss now.

1.3.2 CMB Polarization

Analysing the polarization of the CMB [44] is a complicated field, both from a theoretical and experimental point of view. For a review, see e.g. [45]. Here I want to discuss the main theoretical aspects of polarization, and describe the B-mode polarization that provides a direct signature of gravitational waves [46, 47].

Recall that electromagnetic waves are transverse, which means that the direction of the field oscillation (which determines the polarization) is orthogonal to the propagation of the wave. The intensity (amplitude) of the wave will therefore only vary in a plane orthogonal to the wave vector **k**. If the intensity is the same in any two orthogonal directions in this plane, the wave is unpolarized, otherwise it is polarized. The most general type of polarization is called elliptical, however there are two special cases, circularly polarized waves (where the field amplitude vector traces out a circle in the plane of oscillation) and linearly polarized waves (where the field vector traces out a line) [48].

Before recombination, there is no reason for the photon background to be polarized. However, upon decoupling, the photons' mean free path increases and as long as the photon distribution $\Theta_l(\mathbf{x}, t)$ has a non-zero quadrupole moment ($l = 2$), the wave becomes linearly polarized due to Thomson scattering with electrons [49]. Let me explain this heuristically. Thomson scattering describes how the electric field of the incoming wave excites the electron, which then emits a wave at the same frequency in a different direction. The wave can only retain polarization transverse to the outgoing direction, and will therefore not transmit any intensity in the field component parallel to it, turning an initially unpolarized into a linearly polarized wave.

In the case of the CMB, we do not deal with single plane waves but with a background of photons, which scatter off electrons coming from all directions. It turns out that a background that is either isotropic or only has dipole anisotropy will not be polarized by Thomson scattering [25].

For simplicity, first consider two unpolarized light waves with equal intensity coming from orthogonal directions $\hat{x}$ and $\hat{y}$, and scattering off an electron at the origin that transmits radiation in the $\hat{z}$ direction. The wave propagating towards the electron in the $-\hat{x}$ direction will retain polarization in the $\hat{y}$ direction after scattering, whereas the one from $-\hat{y}$ will keep the $\hat{x}$ component of polarization. Therefore, the transmitted intensity is the same in both directions and the background remains unpolarized. We arrive at the same conclusion for a background with dipole anisotropies, as photons coming from opposite directions with different temperatures will average out. To produce linear polarization, we need a quadrupole moment in the photon distribution. In this case, orthogonal directions are fundamentally different, and therefore the

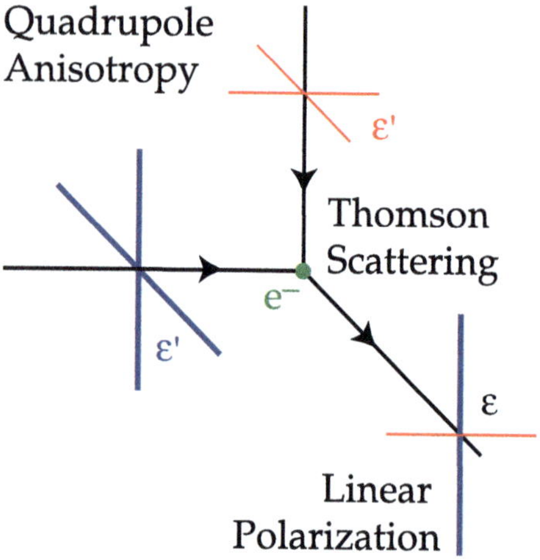

Fig. 1.1 A photon distribution with a quadrupole anisotropy can be linearly polarized from Thomson scattering with electrons. Reprinted from [45], with permission from Elsevier

transmitted intensity will have a preferred direction and the background becomes polarized, see Fig. 1.1.

No further polarization will be induced after the photons have completely decoupled (except for late time polarization when the universe becomes reionized, which leads to a reionization bump in the polarization power spectrum on large scales [50]).

Note that the polarization of the CMB is not very strong, only of the order of a few percent [45], as the quadrupole during the tight coupling regime is small [25]: Before decoupling, a photon's mean free path is very short, so all the photons arriving at a point $\mathbf{x}$ scattered from somewhere nearby. Therefore, they will all have nearly the same temperature, and there is no strong directional dependence in the photon distribution. This corresponds to a monopole perturbation $\Theta_0(\mathbf{x}, t)$ (which is the average over all directions). There is also a significant dipole contribution $\Theta_1(\mathbf{x}, t)$ as the electrons have a bulk velocity and the photons move with them. Monopole and dipole of the fractional temperature distribution therefore dominate, however the quadrupole is still big enough to lead to a measurable polarization signal.

It is straightforward to see why we need a quadrupole moment in the photon distribution to produce polarization when we describe polarization mathematically in terms of Stokes parameters [48]. A polarized wave can be described by the intensity tensor (with a basis of polarization vectors $\hat{\epsilon}_1, \hat{\epsilon}_2$) $I_{ij} = 2\langle \mathbf{E}\mathbf{E}^\dagger \rangle$, where $\mathbf{E}$ is the electric field vector of the polarized wave. Assuming it is moving in the $\hat{z}$ direction,

$$\mathbf{E}(t, z) = A_1 e^{i\phi_1} e^{i(kz-\omega t)} \hat{\epsilon}_1 + A_2 e^{i\phi_2} e^{i(kz-\omega t)} \hat{\epsilon}_2. \tag{1.79}$$

The Hermitian matrix I_{ij} can be expanded in terms of Pauli matrices and written as [25]

$$I_{ij} = \begin{pmatrix} T + Q & U - iV \\ U + iV & T - Q \end{pmatrix}, \tag{1.80}$$

where the coefficients are the Stokes parameters [20]:

$$T = \langle A_1^2 + A_2^2 \rangle \,, \tag{1.81}$$

$$Q = \langle A_1^2 - A_2^2 \rangle \,, \tag{1.82}$$

$$U = \langle 2A_1 A_2 \cos(\phi_1 - \phi_2) \rangle \,, \tag{1.83}$$

$$V = \langle 2A_1 A_2 \sin(\phi_1 - \phi_2) \rangle \,. \tag{1.84}$$

The Stokes parameters satisfy $T^2 = Q^2 + U^2 + V^2$, where T is the total intensity (corresponding to the temperature), Q and U characterise linear polarization (with magnitude $P = \sqrt{Q^2 + U^2}$, angle $\alpha = \frac{1}{2}\tan(U/Q)$) and V describes the degree of circular polarization and is zero for linearly polarized waves (which have field components that are in phase). When we perform a rotation by an angle ψ in the polarization plane, I and V are invariant (scalars) but Q and U transform like a spin-2 field (rotate by an angle 2ψ) [29]:

$$\begin{pmatrix} Q \\ U \end{pmatrix} \rightarrow \begin{pmatrix} \cos 2\psi & \sin 2\psi \\ -\sin 2\psi & \cos 2\psi \end{pmatrix} \begin{pmatrix} Q \\ U \end{pmatrix} \quad \text{or} \quad Q \pm iU \rightarrow e^{2i\psi}[Q \pm iU] \,. \tag{1.85}$$

To produce linear polarization, we therefore need an object with the same transformation properties as the spin-2 field (1.85). The quadrupole of the photon distribution $\Theta_2(\mathbf{k})$ depends on $P_2(\cos\theta) \equiv Y_{20}(\theta)$ from Eq. (1.76), i.e. the second order spherical harmonic, which transforms as spin-2 [51]. It is therefore a necessary requirement for the quadrupole moment of the photon distribution to be non-zero in order to produce polarization from Thomson scattering.

To study what the strength of polarization is today, we need to integrate over all incoming directions at each scattering location and consider all outgoing directions. This calculation requires full use of the Boltzmann equations [25] and can be found in [52–54]. As for the temperature fluctuations, we can define a polarization power spectrum which shows the amount of polarization on different angular scales. As the Q, U parameters are a spin-2 field, we cannot just use ordinary spherical harmonics, but need to revert to spin-weighted spherical harmonics ${}_{\pm 2}Y_{lm}(\hat{n})$ [52]. The field in direction $\hat{n}$ can then be expanded as:

$$[Q \pm iU](\hat{n}) = \sum_{lm} a_{\pm 2,lm} \, {}_{\pm 2}Y_{lm}(\hat{n}) \,. \tag{1.86}$$

Clearly, it would be preferable to describe polarization in terms of scalar quantities, just like we do for $\delta T/T$. Indeed, we can perform a change of basis and define spherical harmonic coefficients for two scalar quantities $E(\hat{n})$ and $B(\hat{n})$ [52, 53]:

$$a_{E,lm} \equiv -\frac{1}{2}\left(a_{2,lm} + a_{-2,lm}\right), \quad a_{B,lm} \equiv -\frac{1}{2i}\left(a_{2,lm} - a_{-2,lm}\right) \,. \tag{1.87}$$

E-modes correspond to polarization fields whose strength varies parallel or perpendicular to the polarization direction (like a curl-free electric field), whereas for B-modes the variation is at 45° (like a divergence-free magnetic field). As scalars, they are invariant under rotations, but only E is invariant under a parity transformation while B changes sign.

The E/B decomposition is useful as you can show that scalars produce only E-modes, whereas tensors produce both E and B [46, 47]. Heuristically, the reason scalars and tensors produce different polarization types can be understood as follows (as explained in [25]): Scalar perturbations can be described by plane waves, where the wave vector $\mathbf{k}$ determines the direction of propagation. There should be rotational symmetry around this wave vector, and only the parity invariant E-modes should be produced. Tensor perturbations, on the other hand, introduce an azimuthal dependence into the photon distribution. This additional component affects the polarization and we find that they can also give rise to a B-mode pattern, which changes sign under a parity transformation.

The fields T, B and E completely describe the photon field and we can define power spectra for each of them. We can use polarization alongside temperature measurements to probe the CMB anisotropies. Unlike the temperature fluctuations that interact with gravitational fields, the polarization pattern does not change after production (except due to lensing [20]), as it can only be generated by scattering.

To describe the photon distribution statistically, we can use autocorrelations of the three different fields, TT, EE and BB, and cross correlations. However, the correlators TB, EB vanish by symmetry arguments (as B is odd under parity) and only the TE cross power spectrum is non-zero [29]. The different angular power spectra can be denoted as

$$C_l^{XY} = \frac{1}{2l+1}\sum_m \langle a^*_{X,lm} a_{Y,lm}\rangle \quad X, Y = T, E, B\,. \tag{1.88}$$

Note that there is a way to produce a non-zero TB and EB power spectrum: If gravity was chiral, i.e. the two tensor polarizations (see Sect. 1.5.1) were different, parity would be violated and the TB and EB correlators would not vanish [55]. It could be easier to detect tensor modes through a measurement of the TB rather than the BB power spectrum, as the amplitude of T is much bigger than that of B. A chiral graviton from loop quantum gravity and its implication for the tensor power spectrum are the subject of the work presented in Chap. 2.

Like the temperature power spectrum, the TE and EE spectra are both dominated by the contribution from scalar modes [29]. Measuring the B-mode power spectrum, on the other hand, enables us to directly probe tensor perturbations. While the E-mode polarization has already been detected around 10 years ago [14, 56], due to the small value of r it is much more difficult to measure B-modes. They might have finally been detected by BICEP2 [17], suggesting a value of $r \approx 0.2$. This result will need to be complemented by a full sky analysis of polarization, as performed by the

Planck collaboration who are still due to release their results, as well as confirmed by other experiments [57, 58].

Finally, a remark: It can be shown that gravitational lensing can distort an E-mode into a B-mode pattern, with the effect peaking on scales $l \sim 1000$ [59]. This will affect the primordial B-mode spectrum due to tensors, but leaves it unaffected at large scales. Lensing B-modes have recently been discovered for the first time by the South Pole Telescope [60].

1.4 Reheating

Reheating describes the transition from the end of inflation, where the universe is filled with an oscillating homogeneous field, to radiation domination. Except for gravitational waves, the subject of Chap. 3, we do not expect many cosmological observables to have been directly affected by this process, making it one of the least probed stages in the early universe.

For a long time, the detailed dynamics of reheating were not well understood and the decay was described by a perturbative, effective theory. In 1994, Lev Kofman and collaborators [61] developed the theory of preheating, which was studied analytically in great detail and describes the early stages of the transition. In this section, I will summarize the main aspects of the theory of reheating.

At the end of inflation, when the inflaton oscillates around the minimum of its potential, we are faced with a problem: during the de Sitter-like exponential expansion, the number density of all particle species reduced dramatically, as $n \propto a^{-3}$. Therefore, the universe at the end of inflation is empty and cold, with all of the energy stored in the homogeneous inflaton field. Somehow, we must recover the hot Big Bang picture (which states that the universe after inflation should be in thermal equilibrium) and all the particle species within it.

The idea of reheating, first discussed in [30], states that while the inflaton ϕ oscillates about the minimum of its potential, it produces elementary particles. After (almost) all of the inflaton energy has been transferred, the decay products thermalise at the reheating temperature T_r, motivating the name reheating (the prefix "re", however, is very misleading, as there was not necessarily a stage of thermal equilibrium before reheating occurred). The physical mechanism leading to particle production was described as a perturbative decay of inflaton particles (which make up the homogeneous field condensate that drove inflation) into other particle species. This process was studied in detail by [62, 63], where the reheating temperature (providing the initial condition for the hot Big Bang picture) for different models was derived.

However, it was realised that the reheating process described in this manner proceeded very slowly and might never complete in some models, for which a lot of the energy remained stored in the inflaton field [61]. The failure of the theory is related to the fact that during the initial stage of reheating, the oscillating homogeneous inflaton should be regarded as a classical condensate, rather than a collection of single particles. If we think of the inflaton as a classical background, it can source quantum fluctuations in the fields it couples to: The oscillations of ϕ result in parametric resonant behaviour in the field fluctuations, leading to exponential growth of certain momentum bands, and hence very efficient particle production.

This process was introduced in [61] and called "preheating", to highlight that it describes the initial stage of reheating. The model considered was that of an inflaton with a quadratic potential $V(\phi) = m^2\phi^2$, coupled to a scalar field χ. It was found that the parametric resonance starts off as broad (with a large range of amplified momenta) and many χ particles are produced, but eventually the resonance becomes narrow and much less efficient. At some point, the resonance ceases and the perturbative description becomes appropriate for the final stage of the decay process.

The analytical investigation of preheating was developed further in [64], where the expansion of the universe and the backreaction of the created particles was taken into account. Moreover, in [65] the model of massless preheating was studied, where the inflaton potential is quartic, $V(\phi) = \lambda\phi^4$.

In Sect. 1.4.1, I will summarize the main aspects of the perturbative theory of reheating and in 1.4.2 I will explain the physics of preheating, emphasising the difference between narrow and broad resonance. These sections are primarily based on the extremely thorough and well written account of reference [64]. Finally, I will describe massless preheating in 1.4.3. This will be the model under consideration when I describe gravitational wave production from preheating in Chap. 3.

1.4.1 Perturbative Reheating

For an inflaton field with a quadratic potential $V(\phi) = \frac{1}{2}m^2\phi^2$, the homogeneous background satisfies [recalling the equation of motion (1.25)]

$$\ddot{\phi} + 3H\dot{\phi} + m^2\phi = 0 \,. \tag{1.89}$$

During inflation, the friction term in H (which is approximately constant) dominates, and the slow-roll conditions imply $m \ll H$. This is true as long as the field values are sufficently large, $\phi > m_{\rm Pl}$ [remember the Hubble rate is inversely proportional to $m_{\rm Pl}$, see Eq. (1.29)].

Inflation ends when ϕ becomes sub-Planckian and the condition $m \ll H$ is no longer satisfied, such that the mass term dominates over the friction term in Eq. (1.89). Therefore, the inflaton behaves like a harmonic oscillator, with a decaying amplitude $\Phi(t)$ due to the damping term $H\dot{\phi}$:

$$\phi(t) \approx \Phi(t)\sin(mt), \quad \Phi(t) \sim \frac{m_{\rm Pl}}{mt} \,. \tag{1.90}$$

During this period (when averaging over several oscillations), the universe evolves as if dominated by matter, so the energy density of the inflaton field decreases as a^{-3}.

Let us now consider couplings between the inflaton and other particles, specifically, the coupling to a scalar χ and a spinor (fermion) ψ. The scalar field represents any bosons (which could be standard model or hidden sector particles) the inflaton might couple to, including vector or higher spin fields. Not including gauge indices will simplify the calculation dramatically, and should capture the relevant dynamics, which will mainly depend on the coupling strength and the type of interaction (e.g. cubic or quartic). The potential term in the Lagrangian is given by

$$V = V(\phi) + \frac{1}{2}g^2\phi^2\chi^2 + h\bar{\psi}\psi\phi\,. \tag{1.91}$$

The coupling constants g, h must be small, as the dynamics during inflation should be dominated by the inflaton. Small couplings to other fields can be written as radiative corrections to the inflaton potential [66], which makes it possible to find limits on the size of these couplings using the CMB constraints [16]. Typically, the coupling constants need to satisfy $g, h \lesssim 10^{-3}$ [67].

For a quadratic inflaton potential with a minimum at $\phi = \sigma$,

$$V(\phi) = \frac{1}{2}m^2(\phi - \sigma)^2\,, \tag{1.92}$$

we can perform a field redefinition $\phi \rightarrow \phi + \sigma$ in the potential (1.91) such that we obtain the usual quadratic term $\frac{1}{2}m^2\phi^2$ plus an additional cubic interaction term $-g^2\sigma\phi\chi^2$. Considering the inflaton field as a coherent wave of ϕ-particles, there are 3-point interactions between the inflaton and the scalar and spinor fields, i.e. decay processes $\phi \rightarrow \chi\chi$, $\phi \rightarrow \psi\psi$, with cross sections [68]

$$\Gamma(\phi \rightarrow \chi\chi) = \frac{g^4\sigma^2}{8\pi m}\,, \quad \Gamma(\phi \rightarrow \psi\psi) = \frac{h^2 m}{8\pi}\,. \tag{1.93}$$

We can see that perturbative decay is a slow process as the cross sections are proportional to powers of the small coupling constants, g^4 and h^2. The corrections to the inflaton potential can be taken care of by introducing a friction term, given by the total cross section $\Gamma = \Gamma(\phi \rightarrow \chi\chi) + \Gamma(\phi \rightarrow \psi\psi)$, in the equation of motion:

$$\ddot{\phi} + 3H\dot{\phi} + \Gamma\dot{\phi} + m^2\phi = 0\,. \tag{1.94}$$

Reheating ends when the Hubble rate becomes smaller than Γ, which signals that the expansion of the universe has become slow enough for the decay of the inflaton to complete. The relativistic decay products then thermalise and the universe becomes radiation dominated. The energy density at this point is

$$\rho_\phi(t_r) = 3\Gamma^2 m_{\rm Pl}^2 = \frac{\pi^2}{30} g_\star T_r^4 \,, \tag{1.95}$$

where the second equality relates the energy density to the reheating temperature (assuming thermal equilibrium), and g_* is the number of relativistic degrees of freedom [20].

1.4.2 Preheating and Parametric Resonance

Let us approach the problem of inflaton decay from a different point of view, and consider the boson as a quantum field $\hat{\chi}$ interacting with a classical background $\phi(t)$ (I will ignore fermions from now on, as they do not partake in the efficient paramteric resonance, which is related to Bose condensation effects). We can expand the field in Fourier space by using the standard mode expansion in terms of creation and annihilation operators (c.f. Sect. 1.2.4),

$$\chi(t, \mathbf{x}) = \frac{1}{(2\pi)^3} \int d^3k \left(\hat{a}_k \chi_k(t) e^{-i\mathbf{k}\cdot\mathbf{x}} + \hat{a}_k{}^\dagger \chi_k^*(t) e^{i\mathbf{k}\cdot\mathbf{x}} \right) \,, \tag{1.96}$$

where $\mathbf{x}$ and $\mathbf{k}$ are comoving quantities. Due to its interaction with the inflaton, χ acquires a time dependent effective mass term $m_{\rm eff} = g^2\phi^2(t)$, and the equation of motion for its mode functions in an expanding background is given by (writing Eq. (1.55) in terms of coordinate time)

$$\ddot{\chi_k} + 3H\dot{\chi_k} + \left(\frac{k^2}{a^2} + g^2 \Phi(t)^2 \sin^2(mt) \right) \chi_k = 0 \,, \tag{1.97}$$

where the solution for $\phi(t)$, Eq. (1.90), was used. Eq. (1.97) describes an oscillator with a time-dependent frequency that changes periodically. This periodicity is the source of parametric resonance.

Ignoring the expansion of the universe (by setting $a = 1$, which makes the amplitude Φ time independent) and defining a new time variable $z = mt$, we can write (1.97) as the Mathieu equation [69]:

$$\chi_k'' + (A_k - 2q\cos(2z))\,\chi_k = 0 \,, \tag{1.98}$$

where a prime denotes differentiation with respect to z and

$$A_k = \frac{k^2}{m^2} + 2q \,, \quad q = \frac{g^2\Phi^2}{4m^2} \,. \tag{1.99}$$

The properties of the Mathieu equation have been extensively studied [69], and its solutions show parametric resonant behaviour: For certain resonant momentum bands Δk, there exists a solution of (1.98) for which $\chi_{\mathbf{k}}$ grows exponentially,

$$\chi_k = \exp(\mu_k t)\, f(t)\,, \tag{1.100}$$

where μ_k is the Floquet exponent and $f(t)$ is a periodic function. An exponential growth of the mode functions leads to an exponential growth in the occupation number, as

$$n_k \sim |\chi_k|^2 \sim \exp(2\mu_k t)\ . \tag{1.101}$$

Therefore, particles are being produced very efficiently during parametric resonance.

The resonance parameter q determines the structure of the resonant bands as well as the Floquet exponents. There are two very different regimes, narrow resonance for $q \ll 1$ and broad resonance for $q \gg 1$. I will study both cases separately.

1.4.2.1 Narrow Resonance

From the definition of q in (1.99), it is clear that we are in the narrow regime $q \ll 1$ when $g\Phi < m$. It was shown in [70] that in this case, resonance occurs for modes $A_k^{(n)} \approx n^2$, where n is an integer, and the width of each band is of order $\Delta A_k^{(n)} \sim q^n$. Therefore, the first band $A_k^{(1)} = 1 \pm q$ will dominate. The centre of this band corresponds to $k \sim m$, which indicates that two ϕ particles with mass m have decayed to two χ particles of momentum k.

Although this looks like the case of perturbative decay, the actual process is completely different, as the exponential amplification of modes means that the growth rate is directly proportional to the number density of produced particles. Narrow resonance can therefore be seen as a Bose condensation effect [1], for which the production becomes more efficient the more particles have already been produced.

The maximal Floquet exponent, corresponding to the middle of the resonance band (outside the band, μ_k becomes imaginary and therefore the field χ simply oscillates), is given by $\mu_k = q/2$. The smallness of q therefore leads to both a very narrow resonance band and a small amplification exponent. The situation worsens when we take the expansion of the universe into account: The inflaton amplitude Φ then decays, making q even smaller and thus decreasing the width of the bands. Also, modes can get redshifted out of the instability bands and simply oscillate.

Narrow resonance is therefore not an extremely efficient process, and will actually only occur if $qm > 3H + \Gamma$ [64]. Otherwise, there is no resonant behaviour and the decay happens perturbatively as in Sect. 1.4.1. As q decays with time faster than the Hubble rate, narrow resonance will inevitably become inefficient eventually, and thus the final stages of reheating should always be described using perturbative methods.

Numerical simulations show [64] that during narrow resonance, for each oscillation of the inflaton, the growing mode $\chi_k = e^{\mu_k t} f(t)$ also oscillates one time. This is very different to the broad resonance case I will consider now.

1.4.2.2 Broad Resonance

The chaotic inflation model with a quadratic potential actually starts with a period of broad resonance, for which q is very large, corresponding to a large (super-Planckian) initial field amplitude Φ. Broad resonance is a lot more complicated than the narrow case, and I will mainly describe it qualitatively as the proper analytical treatment is very involved, see [64].

Let us initially ignore the expansion of the universe. Solving the Mathieu equation numerically shows that χ_k oscillates many times (during which the occupation number remains constant) for each inflaton oscillation and only increases in amplitude during the short periods when $\phi(t)$ crosses zero [64]. The fast oscillations in $\chi_{\mathbf{k}}$ occur as its effective mass $m_{\rm eff}^{\chi} = g\phi(t)$ is much larger (as long as $\phi(t)$ is not close to zero) than the mass of the inflaton m. Away from $\phi(t) = 0$, the frequency $\omega^2 = k^2 + g^2\phi(t)^2$ therefore changes adiabatically, but when the inflaton approaches zero, this condition ceases to be satisfied, i.e. $\dot{\omega} > \omega^2$. At this point, the occupation number density is not well defined, but stabilises to a higher level after adiabaticity is restored.

During broad resonance, particle production occurs for momenta in the range

$$k^2 \lesssim k_*^2 \equiv gm\Phi\,, \tag{1.102}$$

during the time interval $\Delta t_* \sim k_*^{-1}$ for which the evolution is non-adiabatic. Clearly, the range of amplified momenta is much larger than in the narrow resonance case, making broad resonance a lot more efficient.

If we include the expansion of the universe, the upper bound k_* becomes time-dependent and decreases as $\Phi(t)$ decays. However, at the same time, more physical momenta are redshifted into the resonant bands. More importantly, the expansion changes the character of the resonance overall: As Φ decays, the change of the frequency ω is not simply periodic anymore, which means that $\chi_{\mathbf{k}}$ will have a different phase each time the inflaton crosses zero. As shown in [64], this leads to the process of stochastic resonance, for which the occupation number of the field χ can also decrease after a zero crossing, but still grows overall.

Backreaction effects of the produced χ particles, as well as the rescattering of χ particles which produce inflaton particles, need to be taken account in the full treatment of parametric resonance, see [64] for details. In particular, these effects will determine when broad resonance ends and narrow resonance takes over, after q has become small.

1.4.3 Massless Preheating

I will now describe resonance in the case of massless preheating, where the inflaton has a quartic instead of a quadratic potential, making self-interactions possible. Details about this model can be found in [65], which this section is based on.

The potential for massless preheating is given by

$$V(\phi, \chi) = \frac{1}{4}\lambda\phi^4 + \frac{1}{2}g^2\phi^2\chi^2 \,. \tag{1.103}$$

We will see that the resonance in this case strongly depends on the ratio of coupling constants, g^2/λ.

This theory is particularly interesting as it is scale invariant: as the coupling constants λ and g are dimensionless, there is no physical length scale in the Lagrangian. This is opposed to the case of a quadratic potential, where the mass m is dimensionfull. We will see that, assuming the background behaves like pure radiation, we can arrive at an equation of motion for the scalar field fluctuations $\chi_{\mathbf{k}}$ that is independent of the scale factor, showing that the dynamics do not change as the universe expands. Therefore, we can treat massless preheating like a problem in flat Minkowski space. This simplifies the calculation and means that the characteristic dynamics will remain the same throughout, and the resonance only terminates due to backreaction effects.

1.4.3.1 Background Evolution

For a quartic potential, the equation of motion (1.25) for the homogeneous inflaton field $\phi(t)$ becomes

$$\ddot{\phi} + 3H\dot{\phi} + \lambda\phi^3 = 0 \,. \tag{1.104}$$

After the field amplitude has dropped below $\phi < m_{\mathrm{Pl}}$, the friction term in H becomes subdominant and the inflaton starts oscillating. However, as opposed to the quadratic potential case, the oscillations are not sinusoidal, but given by an elliptic cosine. To see this, we need to make a conformal transformation $\varphi = a\phi$ and use conformal time η (1.13), for which Eq. (1.104) becomes

$$\varphi'' + \lambda\varphi^3 - \frac{a''}{a}\varphi = 0 \,. \tag{1.105}$$

It has been shown [71] that, averaged over many oscillations, a scalar field with a quartic potential behaves like radiation, which implies $a(\eta) \sim \eta$ in conformal time. Therefore, we can ignore the last term in (1.105) and arrive at the equation of motion for a scalar field with a quartic potential in Minkowski spacetime,

$$\varphi'' + \lambda\varphi^3 = 0 \,. \tag{1.106}$$

This can be rewritten as the equation for an elliptic function by defining a dimensionless conformal time variable

$$x \equiv \sqrt{\lambda}\varphi_{\mathrm{i}}\eta \,, \tag{1.107}$$

where $\varphi_{\rm i} \approx \left(\frac{12m_{\rm Pl}^2}{\lambda}\right)^{1/4}$ is the amplitude of the field φ. The equality is approximate as the background evolution is not exactly like in radiation domination, and therefore the amplitude is actually weakly time dependent. Note that the amplitude of the original field ϕ decays as $1/a \sim t^{-1/2}$. The solution of (1.106) is given by

$$\varphi = \varphi_{\rm i} f(x) = \varphi_{\rm i}\,{\rm cn}\left(x - x_0, \frac{1}{\sqrt{2}}\right), \tag{1.108}$$

where $f(x)$ is an elliptic cosine, which is a harmonic expansion in terms of ordinary cosines. The constant x_0 will simply shift the phase of the oscillations and will be ignored from now on. As for massless preheating we do not have simple sinusoidal behaviour, there will be some interesting features in the analytical solution.

1.4.3.2 Fluctuations in the Field χ

The equation of motion of the fluctuations $\chi_{\mathbf{k}}$ is given by

$$\ddot{\chi}_k + 3H\dot{\chi}_k + \left(\frac{k^2}{a^2} + g^2\phi(t)^2\right)\chi_k = 0\,. \tag{1.109}$$

Due to the quartic potential, the oscillating background $\phi(t)$ can also source inflaton fluctuations. Their equation of motion has the same form as (1.109), but with the term g^2 replaced by 3λ:

$$\ddot{\phi}_k + 3H\dot{\phi}_k + \left(\frac{k^2}{a^2} + 3\lambda\phi(t)^2\right)\phi_k = 0\,. \tag{1.110}$$

Rescaling both fields by the scale factor, where $X_{\mathbf{k}} = a\chi_{\mathbf{k}}$, and using the time variable x from before we obtain

$$X_k'' + \left[\kappa^2 + \frac{g^2}{\lambda}{\rm cn}^2\left(x, \frac{1}{\sqrt{2}}\right)\right]X_k = 0\,, \tag{1.111}$$

$$\varphi_k'' + \left[\kappa^2 + 3\,{\rm cn}^2\left(x, \frac{1}{\sqrt{2}}\right)\right]\varphi_k = 0\,, \tag{1.112}$$

where $\kappa = k/\sqrt{\lambda}\varphi_{\rm i}$ is a dimensionless comoving momentum. Like in the massive inflaton case, the fluctuations have an oscillatory mass term, which is now given by a Jacobi cosine. The mode equation for the inflaton fluctuations is identical to the one for the fluctuations $X_{\mathbf{k}}$ with $g^2/\lambda = 3$, so we will not need to consider it separately. Equation (1.111) is known is the Lamé equation [72]. It is determined by only two parameters, the ratio of coupling constants g^2/λ, and the comoving momentum κ.

Note that the scale factor a has dropped out of the equations. This only happened because a universe dominated by a scalar field with a quartic potential behaves approximately like radiation [71], which reduced the equation of motion for the background and fluctuations to one in Minkowski space (after a change of coordinates). This is referred to as conformal invariance of massless preheating by many references [65, 73], but remember that the background evolution is not exactly conformally invariant as the scale factor only satisfies $a \propto t^{1/2}$ to a very good approximation.

Like the Mathieu equation, for any choice of couplings g^2/λ, the Lamé equation has unstable solutions for some range of κ. In these momentum bands, the fluctuations grow exponentially as $\chi_k(\tau) = \exp\left[\mu(k, g^2/\lambda)\tau\right] f(\tau)$, where $f(\tau)$ is a periodic function and $\mu(k, g^2/\lambda)$ is the Floquet exponent quantifying the strength of the resonance.

However, the band structure in this case is a lot more unusual: the strength of the resonance and the location and width of the amplified band $\Delta\kappa$ depends sensitively, and in no way monotonically, on the ratio g^2/λ. As opposed to the broad resonance regime which corresponds to $q > 1$, a higher value of g^2/λ does not necessarily correspond to stronger amplification.

For certain values, given by $g^2/\lambda = n(n+1)/2$, where n is an integer, there is only a finite number of resonant momentum bands (all other values have an infinite number of instability bands, which is the also the case for the Mathieu equation). Specifically, $g^2/\lambda = 1$ ($n = 1$) and $g^2/\lambda = 3$ ($n = 3$) have a single instability band, as all higher ones shrink to nodes at this value. Moreover, long wavelength modes are only amplified if the ratio g^2/λ lies between these special integer values. In this case, the first momentum band extends from $\kappa = 0$ to some maximum value κ_{max}. This is demonstrated for low values of g^2/λ in Fig. 1.2, which shows the resonance chart of the Lamé equation. Shaded regions signify instabilities, and darker colours correspond to a larger exponent μ_κ.

1.4.3.3 Dynamics in Different Regimes

The solutions of the Lamé equation can be written in terms of transcendental Jacobi functions [72]. These are quite complicated, and in [65] it was shown that for the special values $g^2/\lambda = n(n+1)/2$ simple, closed form solutions can be obtained. Other interesting cases are the limits $g^2/\lambda \ll 1$ and $g^2/\lambda \gg 1$. For very small values of g^2/λ, the Lamé equation reduces to the Mathieu equation with a small value of q. Therefore, we are in the narrow resonance regime, with resonance bands that have a very small width. However, the expansion of the universe will not affect the band structure, so the narrow resonance for massless preheating is more efficient than its massive counterpart.

For large values of g^2/λ, we find similar behaviour to that of the broad resonance regime: The fluctuations $\chi_{\mathbf{k}}$ oscillate many times for each inflaton oscillation, and only increase in amplitude when the inflaton crosses zero. Furthermore, for $g^2/\lambda \to \infty$, the width of the resonance band is directly proportional to g^2/λ and the characteristic exponent asymptotically approaches its maximum value $\mu_{\text{max}} \approx 0.2377$.

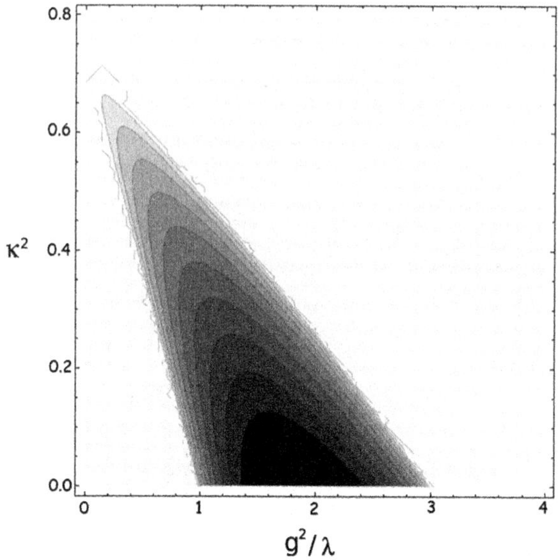

Fig. 1.2 Stability chart of the Lamé equation. Shaded regions correspond to unstable regions where fluctuations grow. The characteristic Floquet exponent $\mu(k, g^2/\lambda)$ is greater for *darker* regions, varying from $\mu \approx 0.2$ (*darkest* region), up to $\mu \approx 0.02$ (*lightest* region), in steps of $\Delta\mu = 0.02$

1.4.3.4 Terminating the Resonance

For massless preheating, due to the disappearance of the scale factor from the equations of motion, the resonance structure is not affected by the expansion of the universe. The only way to terminate the resonance is due to the backreaction of the produced particles (both χ and ϕ), which affect the potential of the oscillating inflaton field. This leads to a restructuring of the resonance bands, and modes that were initially amplified move out of the unstable region and start oscillating. This can happen very fast if the instability band is narrow. The decay of the remaining inflaton condensate then proceeds more slowly again, described by the perturbative decay of Sect. 1.4.1.

1.5 Gravitational Waves

In this section I will give some mathematical background on gravitational waves: I will describe why tensor perturbations correspond to GW degrees of freedom in Sect. 1.5.1, how GWs are generated by sources in 1.5.2 and how much energy they carry in 1.5.3. Finally, in Sect. 1.5.4 I will explain how to detect GWs directly.

I will mainly consider perturbations around flat spacetime. This makes the expressions simpler and captures the important physical aspects of GWs. Note that the Friedmann metric will always look approximately flat on scales much smaller than the Hubble radius H^{-1} for which we can neglect the expansion of the universe [20]. Therefore, when we consider GW produced on subhorizon scales, they will initially behave as if they were in a flat background. However, they will be affected by the expansion of the universe and redshifted [20], so a long time after their production their initial amplitude will have decayed. I will give generalised expressions taking the curved nature of the FRW metric into account when necessary.

1.5.1 Tensor Perturbations as Gravitational Waves

When gravitational fields are weak, we can use the framework of linearised gravity where we write the metric as a fixed background with small perturbations around it. Assuming a flat background for now, we write $g_{\mu\nu} = \eta_{\mu\nu} + h_{\mu\nu}$, where $h_{\mu\nu}$ is a small perturbation. Therefore, we only need to keep terms to linear order in $h_{\mu\nu}$ when we determine the equations of motion from the Einstein equations.

In appendix A.1, I describe how the metric perturbation can be decomposed into scalar, vector and tensor parts, see Eq. (A.1) (where we need to set $a(t) = 1$ for a flat background). In Minkowski space, it can be shown [21] that only the traceless tensor perturbation E_{ij} is a true propagating degree of freedom, while all the others can be derived from it and the stress energy tensor by means of constraint equations. This perturbation carries the degree of freedom corresponding to GWs, but to see its wave nature it is useful to express the perturbed metric in the transverse gauge.

The diffeomorphism invariance of GR requires that physical observables (such as the proper time or the Ricci scalar) do not depend on the choice of coordinates, i.e. they are gauge invariant (see appendix A.1 for more details on selecting gauges in Cosmology). To satisfy this, the full metric perturbation $h_{\mu\nu}$ needs to transform under an infinitesimal change of coordinates $x^\mu \to x^\mu - \xi^\mu$ as [21]

$$h_{\mu\nu} \to h_{\mu\nu} + 2\partial_{(\mu}\xi_{\nu)} \,. \tag{1.113}$$

Choosing a specific transformation ξ^μ determines the gauge and enables us to set certain metric perturbations to zero.

Consider the transverse gauge, for which the choice of ξ^μ ensures that $E_{ij,i} = 0$ and $B_{i,i} = 0$, so both vector and tensor perturbations are transverse [21]. As the tensor perturbation E_{ij} is also traceless by definition, let us denote it as h_{ij}^{TT}, where the superscript TT refers to transverse and traceless. If we consider a situation without a source ($T_{\mu\nu} = 0$), the linearised Einstein equations imply [21] that the metric perturbations in Eq. (A.1) satisfy $\Phi = 0$, $B_i = 0$, $\Psi = 0$ and

$$\Box h_{ij}^{\mathrm{TT}}(\mathbf{x}, t) = 0 \,, \tag{1.114}$$

where $\Box = \partial_\mu \partial^\mu$ is the D'Alembertian operator. You can then simply write the full metric perturbation as

$$h_{\mu\nu}^{\mathrm{TT}} = \begin{pmatrix} 0 & 0 & 0 & 0 \\ 0 & & & \\ 0 & & h_{ij}^{\mathrm{TT}} & \\ 0 & & & \end{pmatrix}, \tag{1.115}$$

where the transverse and traceless tensor $h_{\mu\nu}^{\mathrm{TT}}$ has equation of motion $\Box h_{\mu\nu}^{\mathrm{TT}} = 0$. The solution of this equation is a plane wave, $h_{\mu\nu}^{\mathrm{TT}} = C_{\mu\nu} e^{ik_\alpha x^\alpha}$, which satisfies $k_\mu k^\mu = 0$ (wavevector null) and $k_\mu C^{\mu\nu} = 0$ (wave propagation orthogonal to wave polarization). Therefore, the tensor perturbations h_{ij}^{TT} indeed behave like waves and the first condition shows that they must propagate at the speed of light. For a wave propagating in the z direction, i.e. $k^\mu = (\omega, 0, 0, \omega)$, the second condition implies

$$C_{\mu\nu} = \begin{pmatrix} 0 & 0 & 0 & 0 \\ 0 & h_+ & h_\times & 0 \\ 0 & h_\times & -h_+ & 0 \\ 0 & 0 & 0 & 0 \end{pmatrix}, \tag{1.116}$$

where we defined $h_+ = C_{11}, h_\times = C_{12}$. This shows that GWs have two separate polarizations, denoted by plus and cross due to the way in which they distort test particles. Due to the orthogonality condition $k_\mu C^{\mu\nu} = 0$, test particles will only be perturbed in a direction orthogonal to the propagation direction of the wave. The plus polarization ($h_\times = 0$) perturbs test particles in the same direction they were separated in, whereas the cross polarization perturbs them at 45° to their original separation, see Fig. 1.3.

In an expanding universe, the transverse and traceless tensor metric perturbations are defined as

$$ds^2 = -dt^2 + a^2(t)\left[\delta_{ij} + h_{ij}^{\mathrm{TT}}\right] dx^i dx^j . \tag{1.117}$$

In this case, the wave equation (1.114) acquires a drag term [29]:

$$\ddot{h}_{ij}^{\mathrm{TT}}(\mathbf{x}, t) + 3H\dot{h}_{ij}^{\mathrm{TT}}(\mathbf{x}, t) - \frac{1}{a^2}\nabla^2 h_{ij}^{\mathrm{TT}}(\mathbf{x}, t) = 0 . \tag{1.118}$$

Hence, the amplitude of GWs inside the Hubble volume will decay with time.

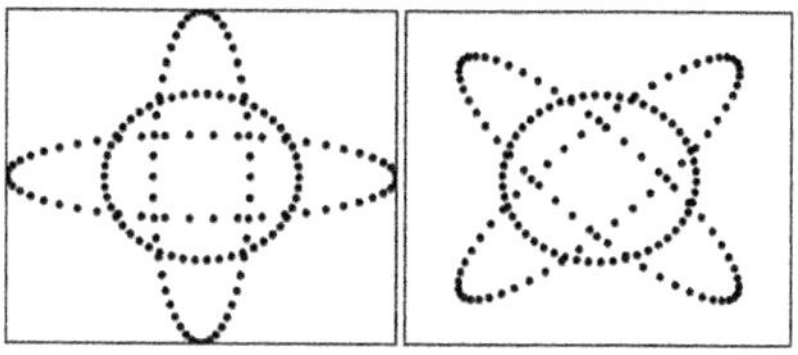

Fig. 1.3 The plus (*left*) and cross (*right*) polarizations of a GW. Figure reproduced from [74]. © http://creativecommons.org/licenses/by-nc-nd/2.0/de/deed.en

1.5.2 Gravitational Waves Generated by Sources

We want to see how to calculate the propagating degrees of freedom of the metric in the presence of sources. If $T_{\mu\nu} \neq 0$, we cannot set Φ, B_i and Ψ to zero. However, there is a useful gauge in this case to describe the behaviour of gravitational waves. Let us first define $\bar{h}_{\mu\nu} = h_{\mu\nu} - \frac{1}{2}h\eta_{\mu\nu}$, leading to a reversed trace, $\bar{h} = -h$. This reduces to the original $h_{\mu\nu}$ in the TT gauge (which can be applied far away from sources).

We will choose the Lorenz gauge in which we can set $\partial_\mu \bar{h}^{\mu\nu} = 0$. Using this condition, the linearised Einstein equation for the trace reversed perturbation is a wave equation for each component with a source term [21],

$$\Box \bar{h}_{\mu\nu} = -16\pi G T_{\mu\nu} . \tag{1.119}$$

To solve this equation, we need to use a Green function $G\,(x^\alpha - y^\alpha)$, which is the solution of the d'Alembertian operator $\Box$ for a delta function source. Equation (1.119) can be expressed as

$$\bar{h}_{\mu\nu} = -16\pi G \int G\left(x^\alpha - y^\alpha\right) T_{\mu\nu}(y) d^4 y , \tag{1.120}$$

where the retarded Green's function (corresponding to waves travelling forward in time) is [21]

$$G\left(x^\alpha - y^\alpha\right) = -\frac{1}{4\pi|\mathbf{x}-\mathbf{y}|}\delta\left[|\mathbf{x}-\mathbf{y}| - (x^0 - y^0)\right]\theta(x^0 - y^0) . \tag{1.121}$$

Plugging Eq. (1.121) into (1.120) and integrating over y^0 gives

$$\bar{h}_{\mu\nu}(t, \mathbf{x}) = 4G \int \frac{1}{|\mathbf{x}-\mathbf{y}|} T_{\mu\nu}(t_r, \mathbf{y}) d^3 y , \tag{1.122}$$

where $t = x^0$ and $t_r = t - |\mathbf{x}-\mathbf{y}|$ is the retarded time. The disturbance for an observer at $(t, \mathbf{x})$ is the sum of contributions at points $(t_r, \mathbf{y})$, where t_r is the coordinate time at which the observers past light cone intersects the source located at a distance $|\mathbf{x}-\mathbf{y}|$.

To solve Eq. (1.122) analytically we need to make a few simplifying assumptions. Consider a situation where we are measuring the wave far away from an isolated, slow-moving source (in this regime, the energy-momentum tensor is negligible, so the tensor perturbations behaves like a propagating wave as in the previous subsection). Note that due to the Lorenz gauge condition we only need to solve for the space-like components of $\bar{h}_{\mu\nu}$ as the components $\bar{h}_{\mu 0}$ can be derived from them. Going through a few steps of algebra (see e.g. [21]), we obtain the quadrupole formula,

$$\bar{h}_{ij}(t, \mathbf{x}) = \frac{2G}{r}\frac{d^2 I_{ij}}{dt^2}(t_r) , \tag{1.123}$$

where the quadrupole moment tensor is given by

$$I_{ij}(t) = \int y^i y^j T^{00}(t, \mathbf{y}) d^3 y \,. \tag{1.124}$$

The gravitational wave produced by an isolated source depends on the second time derivative of the quadrupole tensor (so stationary or spherically symmetric objects would not emit GWs), evaluated at the retarded time.

Compare this to the situation in EM, where electromagnetic radiation is produced by the changing dipole moment of an object. A dipole moment cannot lead to gravitational radiation because of momentum conservation [26]. Therefore we need a quadrupole moment in the source (which measures the shape of the system [21]) to generate GWs. This fact, alongside the general weakness of the coupling of matter to gravity, is why gravitational radiation is so much weaker than its electromagnetic counterpart.

The component of the stress energy-tensor $T_{\mu\nu}$ that is affected by the quadrupole moment of a distribution is the traceless anisotropic stress Π_{ij} [25]. Furthermore, as the GW degrees of freedom are also transverse, only the projection Π_{ij}^{TT} acts as a source term for GWs. In an expanding universe, the wave equation of tensor perturbations can generally be written Eq. (1.118) with a source term, given by the TT part of anisotropic stress [75]:

$$\ddot{h}_{ij}^{\mathrm{TT}}(\mathbf{x}, t) + 3H\dot{h}_{ij}^{\mathrm{TT}}(\mathbf{x}, t) - \frac{1}{a^2}\nabla^2 h_{ij}^{\mathrm{TT}}(\mathbf{x}, t) = 16\pi G \Pi_{ij}^{\mathrm{TT}} \,. \tag{1.125}$$

This expression is valid as long as one can regard the anisotropic stress as a perturbation around a perfect fluid [76].

1.5.2.1 Sources of Gravitational Waves in Cosmology

The production of gravitational waves I have described in this section is a classical process and very different in nature to the primordial gravitational wave background from inflation (see Sect. 1.2.5). This background did not originate from a source, but corresponds to quantum fluctuations in the metric field. These fluctuations were stretched to superhorizon scales during the inflationary phase and result in a scale invariant spectrum of tensor fluctuations.

In this section, I described the emission of GWs by classical sources, due to a time-varying matter distribution with a non-zero quadrupole moment. Gravitational waves are hence produced during many astrophysical phenomena that involve colliding or collapsing bodies, such as binary star systems, coalescing black holes or supernovae (see [74] for a good review on astrophysical GWs). Depending on the details of the system, the emitted GWs from these point sources will peak at specific frequencies.

There could also be stochastic gravitational wave backgrounds (travelling to us from all directions) from the early universe, produced by non-equilibrium phenomena that carry a large amount of energy. Investigating the properties of these backgrounds is an active area of research, which includes bubble collisions during phase transitions [77–81], the creation [82], evolution [83, 84] and decay [85–87] of cosmic defects networks and preheating [88–91], which is the subject of Chap. 3.

1.5.3 Energy Carried by Gravitational Waves

It is natural to ask how much energy is carried by GWs, however calculating this quantity is not easy. Firstly, in GR, there is no local definition of energy as we can always transform to a frame where the perturbation is zero. Even in the case of linearised gravity, it is not obvious what the energy-momentum tensor of the gravitational field should be, as we cannot easily separate gravity from the metric and put it into the right-hand side of Einstein's equation [21].

In order to obtain an expression for the energy carried by GWs, we need to go to higher orders in the expansion. To first order, we cannot feel the effects of gravity, as the Ricci tensor measuring the curvature is zero [26] and test particles therefore move in straight lines. Hence, we need to consider the Einstein equations to second order (this is further motivated by the energy-momentum tensor of EM, which is also quadratic in the fields).

We are interested in the GW energy far away from the source where the vacuum Einstein equations apply. To second order, we can split the Riemann tensors on the LHS of the Einstein equation (1.1) into two parts, one that is linear in the second order perturbation $h^{(2)}_{\mu\nu}$, and one that is quadratic in the first order perturbation $h^{(1)}_{\mu\nu}$. Bringing the second term onto the RHS, we can write this as [21]

$$R^{(1)}_{\mu\nu}[h^{(2)}] - \frac{1}{2}R^{(1)}[h^{(2)}]\eta_{\mu\nu} = -\left(R^{(2)}_{\mu\nu}[h^{(1)}] - \frac{1}{2}R^{(2)}[h^{(1)}]\eta_{\mu\nu}\right). \tag{1.126}$$

If we identify

$$t_{\mu\nu} = -\frac{1}{8\pi G}\left(R^{(2)}_{\mu\nu}[h^{(1)}] - \frac{1}{2}R^{(2)}[h^{(1)}]\eta_{\mu\nu}\right), \tag{1.127}$$

Equation (1.126) is just the Einstein equation for the second order perturbation in the metric, sourced by a gravitational energy-momentum tensor (1.127) that is quadratic in first order perturbations.

This method clearly encodes how the perturbations affect space-time and Eq. (1.127) it is therefore a well motivated choice for the energy-momentum tensor of gravitational waves. Note, however, that $t_{\mu\nu}$ is not a true tensor and, more importantly, it is not gauge invariant [21]. It is however possible to find a gauge-invariant measure of $t_{\mu\nu}$ by averaging over several wavelengths, as this circumvents the non-locality of the description and makes it possible to capture the effects of curvature. In the TT gauge, the averaged energy-momentum tensor can be written as [21]

$$t_{\mu\nu} = \frac{1}{32\pi G}\left\langle (\partial_\mu h^{\mathrm{TT}}_{\alpha\beta})(\partial_\nu h^{\alpha\beta}_{\mathrm{TT}}) \right\rangle . \tag{1.128}$$

The energy density of GWs is simply given by $\rho_{\mathrm{GW}} = t_{00}$. In principle, we can calculate this quantity numerically, by solving Eq. (1.125) for an arbitrary source and deriving the energy-momentum tensor (1.128) from $h^{\mathrm{TT}}_{\mu\nu}$.

1.5.4 Gravitational Wave Detectors

The direct detection of GWs is extremely difficult, and so far we have not been successful in measuring a signal. An indirect detection has been achieved by Hulse and Taylor [22], who observed the change in orbital period of a binary pulsar, which exactly matches the energy loss due to gravitational waves predicted by GR.

Although the flux of energy of GWs can be substantial (which is why Hulse and Taylor were able to observe the energy loss), it is very hard to measure their effect directly [21], as GWs only couple very weakly to matter. As mentioned in Sect. 1.5.1, a passing GW will distort the shape of an object. Due to the coordinate invariance of GR, this is a tidal effect, which cannot be measured locally [92].

A gravitational wave with amplitude h will lead to a fractional change in the size of an object of order

$$\frac{\Delta L}{L} \sim h \,, \tag{1.129}$$

which is called the strain. The maximum amplitude we can expect for typical astrophysical sources is around $h \sim 10^{-21}$ (the smallness of h shows why the linearised gravity approximation works so well). A gravitational wave of this amplitude would result in a minuscule change $\Delta L \sim 10^{-18}$ m over a length of 1 km, which is nine orders of magnitude smaller than the Bohr radius.

To be able to detect such tiny changes, modern GW detectors employ the methods of laser interferometry. A standard interferometer consists of two arms at a 90° angle to each other. Photons entering the tubes will travel through a beamsplitter which sends them down different arms, where they are possibly reflected multiple times, before recombining at a photodiode. The incoming photon beams are in phase, and the interferometer is set up in such a way that there will be no signal unless the outgoing photons are out of phase. A passing GW would stretch one arm and lengthen the other, and could therefore lead to such a phase shift [21].

The Advanced LIGO detector [23], which will start taking data in 2015, consists of two interferometers, each with 4 Km arms, based in Washington and Louisiana. It is necessary to have several detectors to be able to localize GW sources in the sky. LIGO actually collaborates with another experiment, VIRGO in Italy [93], and it is hoped that a new GW detector will be built in India. The spatial configuration of these four detectors would lead to very large sky coverage. GWs could therefore be measured coming from nearly all directions in space [92].

LIGO will be most sensitive to frequencies of the order of 100 Hz, with amplitudes down to $h \sim 10^{-23}$ [94]. It should be able to detect signals from the (non-spherical) collapse of supernovae and the coalescence of neutron stars or black holes [21]. Its main limitations will be due to the many sources of noise, including photon shot noise (due to the random nature of emission by a laser), thermal noise and seismic noise which is particularly important at low frequencies.

The eLISA project [24], which has been delayed multiple times and is supposed to launch in 2032, is a space based interferometer, consisting of three spacecraft carrying test masses and which are arranged as an equilateral triangle. The lengths of the arms is very large, 5 million km, and eLISA would therefore be sensitive to much lower frequencies, between 10^{-4} and 1 Hz, with amplitudes as low as $h \sim 10^{-24}$ [94]. As it is in space, it does not suffer from seismic noise, but will have an additional error source due to inaccuracies in the arm length [21]. Low frequency sources of GWs include certain binary systems and supermassive black holes [21].

While all current direct detectors are mainly aimed at measuring GWs from astrophysical sources, there are many interesting sources of cosmological origin, as mentioned at the end of Sect. 1.5.2. The scale invariant background from inflation is distributed over a vast frequency range, however its amplitude is at least five orders of magnitude lower than the sensitivity of eLISA or LIGO [94]. A proposed space-based detector that might be able to measure the primordial GW background directly in the future is the Big Bang Observer (BBO) [95], which is a configuration of four eLISA type detectors. It would be sensitive to a frequency range between eLISA and LIGO, where no strong signals are expected from astrophysical sources. So far, however, it is much easier to investigate the tensor modes from inflation through the B-mode polarization [17].

For a first order phase transition happening at the electroweak scales, GWs are produced in a range that might be detectable by eLISA [96]. Gravitational waves from preheating, on the other hand, are produced at much higher frequencies and are therefore not accessible by the current detectors [89]. High frequency detectors have recently been proposed [97–99], however their sensitivity might not be sufficient to detect cosmological GW backgrounds.

References

1. V. Mukhanov, *Physical Foundations of Cosmology* (Cambridge University Press, 2005)
2. A. Sakharov, Violation of CP invariance, c asymmetry, and Baryon asymmetry of the universe. Pisma Zh. Eksp. Teor. Fiz. **5**, 32–35 (1967)
3. V. Rubin, N. Thonnard, W.K.J. Ford, Rotational properties of 21 SC galaxies with a large range of luminosities and radii, from NGC 4605 /R = 4kpc/ to UGC 2885 /R = 122 kpc/. Astrophys. J. **238**, 471 (1980)
4. Supernova Search Team Collaboration, A. G. Riess et al., Observational evidence from supernovae for an accelerating universe and a cosmological constant. Astron. J. **116**, 1009–1038 (1998). http://xxx.lanl.gov/abs/astro-ph/9805201

5. Supernova Cosmology Project Collaboration, S. Perlmutter et al., Measurements of Omega and Lambda from 42 high redshift supernovae. Astrophys. J. **517**, 565–586 (1999). http://xxx.lanl.gov/abs/astro-ph/9812133
6. A.A. Starobinsky, Spectrum of relict gravitational radiation and the early state of the universe. JETP Lett. **30**, 682–685 (1979)
7. K. Sato, First order phase transition of a vacuum and expansion of the universe. Mon. Not. Roy. Astron. Soc. **195**, 467–479 (1981)
8. D. Kazanas, Dynamics of the universe and spontaneous symmetry breaking. Astrophys. J. **241**, L59–L63 (1980)
9. A.H. Guth, The inflationary universe: a possible solution to the horizon and flatness problems. Phys. Rev. D **23**, 347–356 (1981)
10. A.A. Penzias, R.W. Wilson, A measurement of excess antenna temperature at 4080-Mc/s. Astrophys. J. **142**, 419–421 (1965)
11. P. Peebles, R. Dicke, Origin of the globular star clusters. Astrophys. J. **154**, 891 (1968)
12. J.C. Mather, E. Cheng, D. Cottingham, R. Eplee, D. Fixsen et al., Measurement of the cosmic microwave background spectrum by the COBE FIRAS instrument. Astrophys. J. **420**, 439–444 (1994)
13. Boomerang Collaboration Collaboration, P. de Bernardis et al., A flat universe from high resolution maps of the cosmic microwave background radiation. Nature **404**, 955–959 (2000). http://xxx.lanl.gov/abs/astro-ph/0004404
14. WMAP Collaboration Collaboration, D. Spergel et al., First year Wilkinson microwave anisotropy probe (WMAP) observations: determination of cosmological parameters. Astrophys. J. Suppl. **148**, 175–194 (2003). http://xxx.lanl.gov/abs/astro-ph/0302209
15. Planck Collaboration, P. Ade et al., Planck 2013 results. I. Overview of products and scientific results. http://xxx.lanl.gov/abs/1303.5062
16. Planck Collaboration Collaboration, P. Ade et al., Planck 2013 results. XXII. Constraints on inflation. http://xxx.lanl.gov/abs/1303.5082
17. BICEP2 Collaboration Collaboration, P.A.R. Ade et al., BICEP2 I: Detection Of B-mode polarization at degree angular scales. http://xxx.lanl.gov/abs/1403.3985
18. J. Cowen, Gravitational waves discovery now officially dead. Nature (2015)
19. R.M. Wald, *General Relativity* (The University of Chicago Press, Chicago, 1984)
20. A.R. Liddle, D.H. Lyth, Cosmological inflation and large-scale structure
21. S. Carroll, *Spacetime and Geometry: An Introduction to General Relativity* (Benjamin Cummings, New York, 2003)
22. R. Hulse, J. Taylor, Discovery of a pulsar in a binary system. Astrophys. J. **195**, L51–L53 (1975)
23. LIGO Scientific Collaboration Collaboration, G. M. Harry, Advanced LIGO: The next generation of gravitational wave detectors. Class. Quant. Grav. **27**, 084006 (2010)
24. P. Amaro-Seoane, S. Aoudia, S. Babak, P. Binetruy, E. Berti, et al., eLISA/NGO: Astrophysics and cosmology in the gravitational-wave millihertz regime. GW Notes **6**, 4–110 (2013). http://xxx.lanl.gov/abs/1201.3621
25. S. Dodelson, *Modern Cosmology* (Academic Press, Elsevier, 2003)
26. J.A. Peacock, *Cosmological Physics* (Cambridge University Press, Cambridge, 2001)
27. E. Noether, Invariant Variation Problems. Gott. Nachr. 235–257 (1918). http://xxx.lanl.gov/abs/physics/0503066
28. V. Mukhanov, S. Winitzki, *Physical Foundations of Cosmology* (Cambridge University Press, Cambridge, 2007)
29. D. Baumann, TASI lectures on inflation. http://xxx.lanl.gov/abs/0907.5424
30. A.D. Linde, A new inflationary universe scenario: a possible solution of the horizon, flatness, homogeneity, isotropy and primordial monopole problems. Phys. Lett. B **108**, 389–393 (1982)
31. A. Albrecht, P.J. Steinhardt, Cosmology for grand unified theories with radiatively induced symmetry breaking. Phys. Rev. Lett. **48**, 1220–1223 (1982)
32. S. Hawking, I. Moss, Supercooled phase transitions in the very early universe. Phys. Lett. B **110**, 35 (1982)

33. A.R. Liddle, D.H. Lyth, COBE, gravitational waves, inflation and extended inflation. Phys. Lett. B **291**, 391–398 (1992). http://xxx.lanl.gov/abs/astro-ph/9208007
34. W. de Sitter, Einstein's theory of gravitation and its astronomical consequences, third paper. Mon. Not. Roy. Astron. Soc. **78**, 3–28 (1917)
35. L. Alabidi, D.H. Lyth, Inflation models and observation. JCAP **0605**, 016 (2006). http://xxx.lanl.gov/abs/astro-ph/0510441
36. A.D. Linde, Hybrid inflation. Phys. Rev. D **49**, 748–754 (1994). http://xxx.lanl.gov/abs/astro-ph/9307002
37. D.H. Lyth, A. Riotto, Particle physics models of inflation and the cosmological density perturbation. Phys. Rept. **314**, 1–146 (1999). http://xxx.lanl.gov/abs/hep-ph/9807278
38. P. Dirac, *Principles of Quantum Mechanics* (Oxford University Press, Oxford, 1982)
39. A.I.M. Rae, *Quantum Mechanics* (CRC Press, New York, 2007)
40. T. Bunch, P. Davies, Quantum field theory in de sitter space: renormalization by point splitting. Proc. Roy. Soc. Lond. **A360**, 117–134 (1978)
41. T.K. Misner, C.J. Wheeler, Gravitation
42. M. Schlosshauer, *Decoherence and the Quantum-to-Classical Transition* (Springer, Berlin, 2008)
43. D. Wands, K.A. Malik, D.H. Lyth, A.R. Liddle, A New approach to the evolution of cosmological perturbations on large scales. Phys. Rev. D **62**, 043527 (2000). http://xxx.lanl.gov/abs/astro-ph/0003278
44. J. Bond, G. Efstathiou, Cosmic background radiation anisotropies in universes dominated by nonbaryonic dark matter. Astrophys. J. **285**, L45–L48 (1984)
45. W. Hu, M.J. White, A CMB polarization primer. New Astron. **2**, 323 (1997). http://xxx.lanl.gov/abs/astro-ph/9706147
46. M. Kamionkowski, A. Kosowsky, A. Stebbins, A Probe of primordial gravity waves and vorticity. Phys. Rev. Lett. **78**, 2058–2061 (1997). http://xxx.lanl.gov/abs/astro-ph/9609132
47. U. Seljak, M. Zaldarriaga, Signature of gravity waves in polarization of the microwave background. Phys. Rev. Lett. **78**, 2054–2057 (1997). http://xxx.lanl.gov/abs/astro-ph/9609169
48. J.D. Jackson, *Classical Electrodynamics*
49. S. Chandrasekhar, *Radiative Transfer* (Dover Publications Inc., New York, 1960)
50. M. Zaldarriaga, Polarization of the microwave background in reionized models. Phys. Rev. D **55**, 1822–1829 (1997). http://xxx.lanl.gov/abs/astro-ph/9608050
51. R. Courant, D. Hilbert, *Methods of Mathematical Physics*, Vol. I. (Wiley-Interscience, New York, 1962)
52. M. Kamionkowski, A. Kosowsky, A. Stebbins, Statistics of cosmic microwave background polarization. Phys. Rev. D **55**, 7368–7388 (1997). http://xxx.lanl.gov/abs/astro-ph/9611125
53. M. Zaldarriaga, U. Seljak, An all sky analysis of polarization in the microwave background. Phys. Rev. D **55**, 1830–1840 (1997). http://xxx.lanl.gov/abs/astro-ph/9609170
54. W. Hu, M.J. White, CMB anisotropies: Total angular momentum method. Phys. Rev. D **56**, 596–615 (1997). http://xxx.lanl.gov/abs/astro-ph/9702170
55. C.R. Contaldi, J. Magueijo, L. Smolin, Anomalous CMB polarization and gravitational chirality. Phys. Rev. Lett. **101**, 141101 (2008). http://xxx.lanl.gov/abs/0806.3082
56. J. Kovac, E. Leitch, C. Pryke, J. Carlstrom, N. Halverson, et al., Detection of polarization in the cosmic microwave background using DASI. Nature **420**, 772–787 (2002). http://xxx.lanl.gov/abs/astro-ph/0209478
57. J. Errard, The new generation CMB B-mode polarization experiment: POLARBEAR. http://xxx.lanl.gov/abs/1011.0763
58. B. Crill, P. Ade, E. Battistelli, S. Benton, R. Bihary, et al., SPIDER: A Balloon-borne Large-scale CMB Polarimeter. http://xxx.lanl.gov/abs/0807.1548
59. M. Zaldarriaga, U. Seljak, Gravitational lensing effect on cosmic microwave background polarization. Phys. Rev. D **58**, 023003 (1998). http://xxx.lanl.gov/abs/astro-ph/9803150
60. SPTpol Collaboration Collaboration, D. Hanson et al., Detection of B-mode Polarization in the cosmic microwave background with data from the south pole telescope. Phys. Rev. Lett. **111**, 141301 (2013). http://xxx.lanl.gov/abs/1307.5830

61. L. Kofman, A.D. Linde, A.A. Starobinsky, Reheating after inflation. Phys. Rev. Lett. **73**, 3195–3198 (1994). http://xxx.lanl.gov/abs/hep-th/9405187
62. A. Dolgov, A.D. Linde, Baryon asymmetry in inflationary universe. Phys. Lett. B **116**, 329 (1982)
63. L. Abbott, E. Farhi, M.B. Wise, Particle production in the new inflationary cosmology. Phys. Lett. B **117**, 29 (1982)
64. L. Kofman, A.D. Linde, A.A. Starobinsky, Towards the theory of reheating after inflation. Phys. Rev. D **56**, 3258–3295 (1997). http://xxx.lanl.gov/abs/hep-ph/9704452
65. P.B. Greene, L. Kofman, A.D. Linde, A.A. Starobinsky, Structure of resonance in preheating after inflation. Phys. Rev. D **56**, 6175–6192 (1997). http://xxx.lanl.gov/abs/hep-ph/9705347
66. M.E. Peskin, D.V. Schroeder, *An Introduction to Quantum Field Theory* (Westview Press Inc., 1995)
67. D.G. Figueroa, Phenomenological and theoretical aspects of reheating. PhD thesis
68. A.D. Linde, *Particle Physics and Inflationary Cosmology* (Harwood, Chur, 1990)
69. N.W. Mac Lachlan, *Theory and Applications of Mathieu Functions*
70. L. Landau, L. Lifshits, *Mechanicss*
71. M.S. Turner, Coherent scalar field oscillations in an expanding universe. Phys. Rev. D **28**, 1243 (1983)
72. A. e. Erdelyi, *Higher Transcendental Functions* (Bateman Manuscript Project), Vol. 3. (McGraw-Hill, New York, 1955)
73. D.I. Kaiser, Resonance structure for preheating with massless fields. Phys. Rev. D **57**, 702–711 (1998). http://xxx.lanl.gov/abs/hep-ph/9707516
74. B. Sathyaprakash, B. Schutz, Physics, astrophysics and cosmology with gravitational waves. Living Rev. Rel. **12**, 2 (2009). http://xxx.lanl.gov/abs/0903.0338
75. V.F. Mukhanov, H. Feldman, R.H. Brandenberger, Theory of cosmological perturbations. Part 1. Classical perturbations. Part 2. Quantum theory of perturbations. Part 3. Extensions. Phys. Rept. **215**, 203–333 (1992)
76. S. Weinberg, *Cosmology* (Oxford University Press, Oxford, 2008)
77. M. Kamionkowski, A. Kosowsky, M.S. Turner, Gravitational radiation from first order phase transitions. Phys. Rev. D **49**, 2837–2851 (1994). http://xxx.lanl.gov/abs/astro-ph/9310044
78. S.J. Huber, T. Konstandin, Gravitational wave production by collisions: more bubbles. JCAP **0809**, 022 (2008). http://xxx.lanl.gov/abs/0806.1828
79. C. Caprini, R. Durrer, G. Servant, The stochastic gravitational wave background from turbulence and magnetic fields generated by a first-order phase transition. JCAP **0912**, 024 (2009). http://xxx.lanl.gov/abs/0909.0622
80. C. Caprini, R. Durrer, T. Konstandin, G. Servant, General properties of the gravitational wave spectrum from phase transitions. Phys. Rev. D **79**, 083519 (2009). http://xxx.lanl.gov/abs/0901.1661
81. M. Hindmarsh, S.J. Huber, K. Rummukainen, D.J. Weir, Gravitational waves from the sound of a first order phase transition. Phys. Rev. Lett. **112**, 041301 (2014). http://xxx.lanl.gov/abs/1304.2433
82. J.-F. Dufaux, D.G. Figueroa, J. Garcia-Bellido, Gravitational waves from Abelian Gauge fields and cosmic strings at preheating. Phys. Rev. D **82**, 083518 (2010). http://xxx.lanl.gov/abs/1006.0217
83. E. Fenu, D.G. Figueroa, R. Durrer, J. Garcia-Bellido, Gravitational waves from self-ordering scalar fields. JCAP **0910**, 005 (2009). http://xxx.lanl.gov/abs/0908.0425
84. D.G. Figueroa, M. Hindmarsh, J. Urrestilla, Exact scale-invariant background of gravitational waves from cosmic defects. Phys. Rev. Lett. **110**(10), 101302 (2013). http://xxx.lanl.gov/abs/1212.5458
85. A. Vilenkin, Gravitational radiation from cosmic strings. Phys. Lett. B **107**, 47–50 (1981)
86. T. Vachaspati, A. Vilenkin, Gravitational radiation from cosmic strings. Phys. Rev. D **31**, 3052 (1985)
87. S. Olmez, V. Mandic, X. Siemens, Gravitational-wave stochastic background from kinks and cusps on cosmic strings. Phys. Rev. D **81**, 104028 (2010). http://xxx.lanl.gov/abs/1004.0890

88. S. Khlebnikov, I. Tkachev, Relic gravitational waves produced after preheating. Phys. Rev. D **56**, 653–660 (1997). http://xxx.lanl.gov/abs/hep-ph/9701423
89. R. Easther, E.A. Lim, Stochastic gravitational wave production after inflation. JCAP **0604**, 010 (2006). http://xxx.lanl.gov/abs/astro-ph/0601617
90. J. Garcia-Bellido, D.G. Figueroa, A. Sastre, A gravitational wave background from reheating after hybrid inflation. Phys. Rev. D **77**, 043517 (2008). http://xxx.lanl.gov/abs/0707.0839
91. J.F. Dufaux, A. Bergman, G.N. Felder, L. Kofman, J.-P. Uzan, Theory and numerics of gravitational waves from preheating after inflation. Phys. Rev. D **76**, 123517 (2007). http://xxx.lanl.gov/abs/0707.0875
92. D. Holz, Gravitational wave cosmology. Lectures given at "Essential Cosmology for the Next Generation/Cosmology on the Beach" Conference, Cabo San Lucas, Mexico, 13–17 Jan 2014
93. LIGO Scientific Collaboration, Virgo Collaboration Collaboration, J. Abadie et al., Search for Gravitational Waves from Compact Binary Coalescence in LIGO and Virgo Data from S5 and VSR1. Phys. Rev. D **82**, 102001 (2010). http://xxx.lanl.gov/abs/1005.4655
94. M. Maggiore, Gravitational wave experiments and early universe cosmology. Phys. Rept. **331**, 283–367 (2000). http://xxx.lanl.gov/abs/gr-qc/9909001
95. J. Crowder, N.J. Cornish, Beyond LISA: exploring future gravitational wave missions. Phys. Rev. D **72**, 083005 (2005). http://xxx.lanl.gov/abs/gr-qc/0506015
96. C. Grojean, G. Servant, Gravitational waves from phase transitions at the electroweak scale and beyond. Phys. Rev. D **75**, 043507 (2007). http://xxx.lanl.gov/abs/hep-ph/0607107
97. A. Cruise, R. Ingley, A prototype gravitational wave detector for 100-MHz. Class. Quant. Grav. **23**, 6185–6193 (2006)
98. A. Cruise, The potential for very high-frequency gravitational wave detection. Class. Quant. Grav. **29**, 095003 (2012)
99. T. Akutsu, S. Kawamura, A. Nishizawa, K. Arai, K. Yamamoto, et al., Search for a stochastic background of 100-MHz gravitational waves with laser interferometers. Phys. Rev. Lett. **101**, 101101 (2008). http://xxx.lanl.gov/abs/0803.4094

Chapter 2
Chiral Tensor Power Spectrum from Quantum Gravity

Quantum fluctuations that are produced during inflation freeze out after leaving the horizon and can survive until today, as was described in Sect. 1.2.5. These fluctuations, having been produced in the very early universe, might carry some information about the quantum nature of gravity. The theory of loop quantum gravity does not use the metric as its fundamental gravitational variable, but a (generally) complex connection. Therefore, deriving the power spectrum of tensor perturbations in this framework, which was done in Sect. 1.2.5 in the standard second order formalism, could lead to a different result. Considering new variables to describe spacetime is always interesting from a quantum mechanical point of view, as different quantum theories can give rise to equivalent classical theories [1]. We cannot know from first principles which description is the correct one, and experiments that involve quantum mechanical observables like power spectra might be the only way of finding out.

I will first outline general principles of the canonical quantization of gravity in Sect. 2.1, starting with the usual approach taken in quantum field theory, and then describing the framework of loop quantum gravity. I will finish by comparing the canonical and covariant approaches to quantization.

In the Sect. 2.2, I will describe different formalisms used in general relativity. In particular, the tetrad formalism, the first order formalism which results in the Palatini action, and the Ashtekar formalism which forms the basis of loop quantum gravity will be discussed.

Section 2.3 is based on work that has been published in [2, 3]. I will describe how using the Ashtekar variables instead of the standard metric variables to find a perturbed gravitational action during inflation leads to a chirality in the tensor power spectrum, which could leave an observable signature in the CMB. Even though the Ashtekar variables are motivated by loop quantum gravity, they are interesting to study regardless of the success of the theory. If we were to observe a chiral tensor spectrum, it might not necessarily mean that LQG is the correct description of quantum gravity, but it would definitely give us insight into the quantum nature of spacetime.

© Springer International Publishing Switzerland 2015

L.B. Bethke, *Exploring the Early Universe with Gravitational Waves*,
Springer Theses, DOI 10.1007/978-3-319-17449-5_2

2.1 Canonical Quantization of Gravity

In this section, I will briefly discuss the main aspects of quantum field theory (QFT), especially regarding the quantization of gravity, before giving some background on loop quantum gravity, highlighting its successes and shortcomings. I will finish by stressing why it might be interesting for Cosmology to consider the Ashtekar variables, which are motivated by the canonical theory of LQG, as the fundamental variables describing spacetime.

2.1.1 Quantum Field Theory

Quantum field theory is the union of quantum mechanics and special relativity, where instead of considering single particle states, we consider fields which are quantized over a (typically) flat, Minkowski background [4].

When one first studies QFT as an undergraduate, one probably learns how to quantize a scalar field canonically, i.e. using the formalism of Sect. 1.2.4. The canonical quantization procedure [5] has been very successful in the context of QFT, and is used in particular to build the theory of quantum electrodynamics (QED), which has made experimentally verified predictions with astonishing accuracy [6]. One conceptual problem with the approach is the lack of manifest Lorentz invariance due to the splitting of space and time, although the Feynman rules one derives to describe interactions obey the Lorentz symmetry [4].

An alternative approach to quantization is the path integral formalism [7], which uses the Lorentz invariant Lagrangian as its central dynamical variable. It also preserves all other symmetries of the theory and is therefore more suited to treating non-Abelian gauge theories like quantum chromodynamics (QCD) [6].

Although the two approaches lead to equivalent results, depending on the situation, one might be more suitable than the other [4], although the path integral formalism is usually the method of choice for the most developed theory of quantum gravity to date, string theory [1, 8].

In all realistic field theories, ultraviolet divergences arise; which means that at very high energies certain quantities of interest become infinite. Using the procedure of renormalization, we can deal with these divergences and arrive at a physically meaningful theory [4, 6]. It is a well known fact that this procedure fails in the case of gravity: When one tries to quantize the graviton field by treating it as a perturbation around flat spacetime [4], divergences arise that cannot be renormalized. This is probably not surprising; most field theories are effective in the sense that their regime of validity does not extend to the highest energy scales [4]. For gravity itself, it seems we cannot simply cut off the highest energy modes and ignore the nature of spacetime at the Planck scale.

2.1.2 Loop Quantum Gravity

Loop Quantum gravity is an attempt to find a quantum theory of gravity in the most "conservative" [9] way: Its aim is to quantize gravity in a background independent (as the background itself is quantized), non-perturbative manner, without resorting to new physics like higher dimensions, supersymmetry or trying to arrive at a unified description of all fundamental forces. This is in contrast to string theory, which incorporates all these aspects and is also based on the standard QFT approach of quantizing over a fixed, flat background spacetime. LQG, on the other hand, uses a canonical quantization method.

In LQG, we do not want to consider gravitons propagating on a fixed background as one would do in standard QFT, but rather define operators corresponding to spacetime itself. Therefore, the canonical variables should describe spacetime, and indeed the metric was chosen as the central gravitational variable (with its conjugate being related to the extrinsic curvature) in the first attempt of defining a canonical quantum theory of gravity, the ADM formalism [10].

In all canonical theories of GR we need to satisfy a number of constraints, which correspond to the quantum Einstein equations [11] and incorporate diffeomorphism invariance. Appendix A.2 gives some background on Hamiltonian constrained systems, and the specific constraints arising in LQG are given in Sect. 2.3.2. In particular, the Hamiltonian constraint, which corresponds to invariance under time translations, on a quantum level leads to the Wheeler-DeWitt equation $\mathcal{H}|\Psi\rangle = 0$ [12], where the quantum Hamiltonian $\mathcal{H}$ acts on the "wave function of the universe" $|\Psi\rangle$. It is constrained to vanish to reflect the fact that there is no global time variable in GR (this is simply the analogue of the Schrödinger equation in canonical quantum gravity).

Within the ADM formalism, it was very difficult to solve this constraint with the chosen quantum operators. In 1986, Ashtekhar introduced a set of new variables [13, 14], discussed in Sect. 2.2.3, where the central canonical variable is a connection, and its conjugate a (densitised) metric field. Further work by [15, 16] led to the definition of the loop representation (hence the name LQG): The actual variables promoted to field operators were the holonomy (parallel transport around a closed loop) of the connection, and a flux of the densitised metric [11]. Like the creation and annihilation operators of particle states in Sect. 1.2.4, these operators create and destroy "loop states", quantum excitations of spacetime along a single loop [9] (the idea of a loop basis was also used in the context of Yang Mills theory in terms of the Wilson loop [17]).

This approach greatly simplified solving the constraint equations [18], especially after work by Thiemann [19]. The Hilbert space these loop states live in has a basis in terms of spin network states [20, 21]. It is possible to define area and volume operators acting on these spin networks (which can be regarded as building blocks of spacetime [9]) with discrete spectra [22, 23], showing that spacetime is fundamentally discrete in LQG.

Kinematically, the theory is well developed: There exists a well defined scalar product [24, 25] and matter can be coupled to the theory [26, 27]. Progress has also recently been made on identifying n-point functions [28], and therefore an expression for the graviton propagator can be obtained [29]. However, the dynamics of the theory are still not well understood and the low-energy limit that should yield GR has not been established [30].

LQG also has some applications to other areas of physics. It provides a way to calculate the Bekenstein-Hawking entropy [31] and has also spawned the field of loop quantum Cosmology [32, 33]. Loop quantum Cosmology contains some interesting results, including a possible mechanism for driving inflation [34], the absence of singularities [35] and the replacement of the Big Bang by a Big Bounce [36]. However, the approach I will take below is not comparable; I will only be using the Ashtekhar variables, not the loop representation which is the foundation of the LQG formalism.

2.1.3 Different Approaches in Quantum Gravity

In canonical quantum gravity spacetime has to be foliated into spacelike slices evolving in time to be able to define the canonical variables [10]. This introduces an explicit time dependence which manifestly breaks Lorentz invariance. The initial lack of covariance (invariance under general coordinate transformations) and the related problem of defining dynamics are the main criticisms faced by this approach.

Although a path integral formulation of LQG now exists using spinfoams [37, 38], it is still in its infancy and work remains to be done trying to link the different formalisms [30]. Arguably the most successful attempt at trying to find a fundamental theory of quantum gravity to date is string theory [1, 8], which is a covariant approach and therefore does not suffer from the same problems as LQG (although proponents of the latter theory will claim that on the other hand, string theory does not address the principle of background independence in GR, needing to define a fixed background [9]).

Of course, there are many other approaches to tackling the problem of quantum gravity, for example causal dynamical triangulation [39] (which is similar in nature to the spinfoam formalism) or causal set theory [40, 41], where the causal structure of spacetime is taken as the most important physical ingredient.

While a mathematically consistent theory of quantum gravity would obviously be a major breakthrough in theoretical physics, any consistent theory will suffer from the problem that it seems impossible with current technology to make testable predictions: The energy scales at which quantum gravity effects play a role are far too high to be probed directly by experiment. Indirect evidence seems to be the best we can hope for at the moment, and Cosmology is a great candidate to provide just that. Clearly, the conditions right after the Big Bang were such that quantum gravity effects must have played a central role, and they might have left an imprint in the CMB through inflation, which explicitly describes how quantum fluctuations become classical observables. Deriving the spectrum of tensor perturbations using the Ashtekar formalism would provide a test for the predictive power of the theory.

2.2 Different Formalisms for General Relativity

Usually, the protagonist of GR is the metric $g_{\mu\nu}$, and the dynamics are defined by the Einstein-Hilbert action (1.8). However, we can also describe gravitational degrees of freedom using a formulation in terms of tetrads (which requires introducing the language of differential forms), as described in Sect. 2.2.1. The content of this section is based on section 2.9 and Appendix J of [42]. I will continue by introducing the first order formalism in 2.2.2, where the metric and connection are taken to be independent initially, giving the Palatini action. Combining both of these ingredients makes it possible to define the Ashtekar formalism in Sect. 2.2.3.

2.2.1 The Tetrad Formalism

It is sometimes useful, especially when trying to treat GR as a gauge theory, to use a non-coordinate basis as opposed to the standard basis vectors dx^μ, ∂_μ. Motivated by the fact that you can always define a local inertial frame in GR which looks flat, consider the tetrad basis e^I, $I = 1\ldots 4$, that satisfies $ds^2 = \eta_{IJ} e^I e^J$, where η_{IJ} is the Minkowski metric. I is an "internal" index and transforms under the vector representation of the Lorentz group SO(3,1) [43]. We can write the basis vectors e^I in terms of the old coordinate basis as

$$e^I = e^I{}_\mu dx^\mu \,, \tag{2.1}$$

so the defining condition for the tetrad basis can be written in components as

$$g_{\mu\nu} = \eta_{IJ} e^I{}_\mu e^J{}_\nu \,. \tag{2.2}$$

The spacetime indices, denoted by Greek letters, can be raised and lowered using the metric $g_{\mu\nu}$ and transform by general coordinate transformations, while the internal indices, denoted by capital Latin letters, can be raised and lowered using the Minkowski metric η_{IJ} and transform by local Lorentz transformations. The components satisfy orthogonality conditions,

$$e^I{}_\mu e^\mu{}_J = \delta^I_J \,, \quad e^\mu{}_I e^I{}_\nu = \delta^\mu_\nu \,. \tag{2.3}$$

We can also use the components $e^I{}_\mu$ of the tetrad basis to relate the components of a vector V in each basis:

$$V^I = e^I{}_\mu V^\mu \,. \tag{2.4}$$

To be able to use covariant derivatives in this formalism, we need to define the spin connection $\omega_\mu{}^I{}_J$. The covariant derivative of some tensor $A^I{}_J$ is then given by

$$\nabla_\mu A^I{}_J = \partial_\mu A^I{}_J + \omega_\mu{}^I{}_K A^K{}_J - \omega_\mu{}^K{}_J A^I{}_K \,. \tag{2.5}$$

To obtain the defining relations for the spin connection and the curvature in the tetrad basis, it helps to simplify expressions if we use the language of differential forms. Let me define them and list some of their properties.

A differential p-form is a $(0, p)$ antisymmetric tensor (i.e. a 0-form is a scalar, and a one-form is a dual vector $\omega = \omega_\mu dx^\mu$). The (components of the) wedge product between a p-form A and q-form B is an antisymmetrised tensor product,

$$(A \wedge B)_{\mu_1 \ldots \mu_{p+q}} = \frac{(p+q)!}{p!q!} A_{[\mu_1 \ldots \mu_p} B_{\mu_{p+1} \ldots \mu_{p+q}]} \,. \tag{2.6}$$

A basis for p-forms can be written using the wedge product as $\frac{1}{p!} dx^{\mu_1} \wedge \ldots \wedge dx^{\mu_p}$. A p-form A is then given by

$$A = \frac{1}{p!} A_{\mu_1 \ldots \mu_p} dx^{\mu_1} \wedge \ldots \wedge dx^{\mu_p} \,, \tag{2.7}$$

where the components $A_{\mu_1 \ldots \mu_p}$ are totally antisymmetric.

We will also need the exterior derivative which is an antisymmetrised partial derivative that maps a p-form into a $p+1$-form [42]:

$$(dA)_{\mu_1 \ldots \mu_{p+1}} = (p+1) \partial_{[\mu_1} A_{\mu_2 \ldots \mu_{p+q}]} . \tag{2.8}$$

Specifically, for zero and one-forms, i.e. a scalar ϕ and vector $\omega = \omega_\mu dx^\mu$, the exterior derivative is

$$(d\phi) = \partial_\mu \phi \, dx^\mu \,, \quad (d\omega) = \partial_{[\mu} \omega_{\nu]} dx^\mu \wedge dx^\nu \,. \tag{2.9}$$

Since partial derivatives commute, and the exterior derivative is antisymmetric, we have $d(dA) = d^2 A = 0$ for any p-form A.

An important property of the exterior derivative is its action on the wedge product of two forms. If A is a p-form,

$$d(A \wedge B) = dA \wedge B + (-1)^p A \wedge dB \,. \tag{2.10}$$

Finally, one can use n-forms ω in n dimensions to define integration on the manifold, specifically $\int \omega = \int \omega_{0123} d^n x$. As differential forms are completely antisymmetrised, there is only one independent component for an n-form in n dimensions.

We can write the tetrad basis and the spin-connection as one-forms e^I and $\omega^I{}_J$ by suppressing their spacetime indices. The Cartan equations provide defining relations for the torsion and the Riemann tensor in the tetrad basis:

$$T^I \equiv de^I + \omega^I{}_J \wedge e^J \,, \tag{2.11}$$

$$R^I{}_J \equiv d\omega^I{}_J + \omega^I{}_K \wedge \omega^K{}_J \,. \tag{2.12}$$

Note that $R^I{}_J$ is a two-form; it specifies the entire Riemann tensor (not the Ricci tensor). It can be regarded as the field strength of the spin-connection [43]. The Christoffel connection, Eq. (1.5), that is commonly used in GR is torsion-free and ensures $\nabla_\alpha g_{\mu\nu} = 0$. The first property leads to Eq. (2.11) being zero, which gives a condition for the spin connection in terms of the tetrad, and the second implies that the spin connection must be antisymmetric, $\omega^\mu_{IJ} = -\omega^\mu_{JI}$.

The tetrad formalism actually makes calculating metric components, spin connection and Riemann tensor a lot simpler than the usual coordinate approach. As we will make use of them in Sect. 2.3, I will derive a tetrad basis and the associated spin connection for a flat Friedmann background (see also Appendix J of [42]).

For a flat FRW metric (1.12) using conformal time, we have $g_{\mu\mu} = a^2$ (no sum), with all off-diagonal components zero. We need to satisfy Eq. (2.2), and clearly the choice $e^0{}_0 = e^1{}_1 = e^2{}_2 = e^3{}_3 = a$ does the job (any other choice will be related to this by a local Lorentz transformation [42]). The four tetrad forms $e^I = e^I_\mu dx^\mu$, $I = 0, i$, can then be written as

$$e^0 = a d\eta\,, \tag{2.13}$$

$$e^i = a dx^i\,. \tag{2.14}$$

We can derive the components of the spin connection $\omega^I{}_J$ using the torsion free condition (2.11). First though, due to the antisymmetry $\omega_{IJ} = -\omega_{JI}$, we see that

$$\omega^0{}_0 = 0\,, \tag{2.15}$$

$$\omega^0{}_i = \omega^i{}_0\,, \tag{2.16}$$

$$\omega^i{}_j = -\omega^j{}_i\,, \tag{2.17}$$

where we had to raise and lower indices with the Minkowski metric.

Let us solve Eq. (2.11) separately for the $I = 0$ and $I = i$ components (which all have the same form) using the solutions for the tetrad. To take the exterior derivatives, regard the forms in Eqs. (2.13) and (2.14) as a product of a scalar and a one-form and then use the product rule in Eq. (2.10) to obtain (remembering that $d^2 = 0$) $de^0 = a' d\eta \wedge d\eta = 0$ and $de^i = a' d\eta \wedge dx^i$. For $I = 0$, the torsion free condition then gives

$$a\omega^0{}_i \wedge dx^i = 0\,, \tag{2.18}$$

where I used $\omega^0{}_0 = 0$. For $I = i$, we obtain

$$a' d\eta \wedge dx^i + a\omega^i{}_0 \wedge d\eta + a\omega^i{}_j \wedge dx^j = 0\,. \tag{2.19}$$

The only solution compatible with the antisymmetry of the spin connection is to set $\omega^i{}_j = 0$ as well, with the only non-zero component being $\omega^i{}_0 = (a'/a)dx^i = He^i$ [42]. This clearly solves the torsion free conditions, Eqs. (2.18) and (2.19).

2.2.2 The Palatini Formalism

We can rewrite the Einstein-Hilbert action (1.8) using tetrads (remember integration over 4-forms is well defined in four dimensions). The result is [43]

$$S_{\rm EH}\left(g_{\mu\nu}(e)\right)=\frac{1}{2}\int\epsilon_{IJKL}e^{I}\wedge e^{J}\wedge R^{KL}\left(\omega(e)\right). \tag{2.20}$$

This makes the internal gauge symmetry under local Lorentz transformation more apparent [43]. Now, consider the following change in viewpoint: Instead of thinking of the action as a function of the tetrad e^I only, we can initially regard it as a function of both e^I and $\omega^I{}_J$, and keep metric and connection independent. The resulting action,

$$S_{\rm PK}\left(e^{I}_{\mu},\omega^{IJ}_{\mu}\right)=\frac{1}{2}\int\epsilon_{IJKL}e^{I}\wedge e^{J}\wedge R^{KL}(\omega), \tag{2.21}$$

is called the Palatini-Kibble action [19]. Varying it with respect to the metric gives the usual Einstein equations, and varying with respect to $\omega^I{}_J$ shows that it is indeed the spin connection $\omega(e)$ we defined, i.e. it satisfies the torsion-free Cartan equation (2.11) and it is manifestly antisymmetric. This is also known as the first order formalism [44] as the equations of motion only contain first derivatives of metric and connection, while the second order formalism of the Einstein-Hilbert action contains second derivatives of $g_{\mu\nu}$.

2.2.3 The Ashtekar Formalism

We can make a further generalization of the Palatini action and add a term $\delta_{IJKL}e^I\wedge e^J\wedge R^{KL}(\omega)$, where $\delta_{IJKL}=\delta_{I[K}\delta_{L]J}$. This term is compatible with the symmetries and vanishes on-shell, when we use the equation of motion for the spin connection $\omega(e)$ [43]. This gives the Holst action [45]

$$S_{\rm H}\left(e^{I}_{\mu},\omega^{IJ}_{\mu}\right)=\left(\frac{1}{2}\epsilon_{IJKL}+\frac{1}{\gamma}\delta_{IJKL}\right)\int e^{I}\wedge e^{J}\wedge R^{KL}(\omega), \tag{2.22}$$

where the coupling constant introduces the Immirzi parameter γ. This parameter will not appear in the classical theory; however, it does play a role in the quantum theory as we will show later in Sect. 2.3, and also appears in the black hole entropy formula derived for LQG [31].

The Holst action is the fundamental action of loop quantum gravity and can be used to derive the new set of canonical variables in terms of a connection A and its conjugate E, which is related to the metric. This choice greatly simplified the constraint algebra [30] compared to the old ADM formalism [10]. We can write Eq. (2.22) as [43]

$$S(A, E, N, N^a) = \frac{m_{\rm Pl}^2}{\gamma} \int d^4x \left[\dot{A}^i_a E^a_i - A^i_0 G_i - NH - N^a H_a \right] , \tag{2.23}$$

where (A, E) are the canonically conjugated variables, and A^i_0, N and N^a are Lagrange multipliers for the first class constraints (see appendix A.2 for details on constrained systems). The Hamiltonian H and the space diffeomorphism constraint H^a encode the invariance of the action under time translations and spatial diffeomorphisms, and the Gauss constraint G_i generates SU(2) gauge transformations. As we have made a specific choice for the time coordinate, the local Lorentz symmetry is broken to a local SO(3)$\sim$SU(2) symmetry transforming the objects E^a_i and A^i_a.

The canonical variables satisfy commutation relations [43]

$$\{A^i_a(\mathbf{x}), E^b_j(\mathbf{y})\} = \frac{\gamma}{m_{\rm Pl}^2} \delta^b_a \delta^i_j \delta(\mathbf{x} - \mathbf{y}) . \tag{2.24}$$

Specifically, E corresponds to the densitized inverse triad

$$E^a_i = \det \left(e^j_b \right) e^a_i , \tag{2.25}$$

where $i = 1, 2, 3$ is an internal index, and $a = 1, 2, 3$ a spatial index; and A to the SU(2) connection (as opposed to a Lorentz connection) [43]

$$A^i_a = -\frac{1}{2} \epsilon^{ijk} \omega^{jk}{}_a + \gamma \omega^{0i}{}_a , \tag{2.26}$$

where $\omega^I{}_J$ is the spin connection satisfying the torsion-free condition. Defining a mapping (see [19], p.127)

$$\omega^i = -\frac{1}{2} \epsilon^{ijk} \omega^{jk} , \tag{2.27}$$

the connection can also be written as

$$A^i = \omega^i + \gamma \omega^{0i} . \tag{2.28}$$

The original variables chosen by Ashtekar [13] were defined for an Immirzi parameter $\gamma = \pm i$. They are special in the sense that the symmetry group of the connection can be identified with the self dual (SD) SU(2) subgroup of the Lorentz symmetry for $\gamma = i$, and the anti-self dual (ASD) SU(2) for $\gamma = -i$ [43]. These subgroups correspond to the isomorphism between the complexified Lorentz group and SU(2)$\times$SU(2). Hence, I will refer to A^i as the SD connection if $\gamma = i$, and as the ASD connection if $\gamma = -i$.

2.3 Spectrum of Tensor Perturbations Using Ashtekar Variables

In this section I will study the tensor perturbations and calculate their power spectrum within the Ashtekar formalism. First, I will identify the canonical variables, perturbed to first order to describe metric perturbations, in Sect. 2.3.1.

The Hamiltonian description is discussed in 2.3.2: The constraints arising in the Ashtekar formalism will be discussed and Hamilton's equations will be derived for the full and the perturbed variables. Finally, I will derive the second order Hamiltonian describing the dynamics of gravitons (and therefore encoding tensor perturbations). Although classically it reduces to the well-known result presented in Sect. 1.2.5, it is still very instructive to carry out the calculation explicitly as a number of subtleties need to be taken into account which had not been previously identified in the literature.

In Sect. 2.3.3, I will expand the perturbation variables in Fourier space. As the connection is complex, there will be separate positive and negative frequency modes corresponding to gravitons and anti-gravitons, which are related by reality conditions. I will end the section by deriving the commutation relations for the modes.

The quantum theory can then be discussed in Sect. 2.3.4. The Fourier space Hamiltonian can be written in terms of graviton creation and annihilation operators which are linear combinations of the metric and connection. Having identified these operators, we can set up a Hilbert space of graviton states. The states with negative energy are not normalisable under the chosen inner product, which is fixed by the reality conditions. Therefore, half of the graviton operators are unphysical and should be removed, after which we are left with the usual two graviton polarizations. I will show that after normal ordering, we obtain a chiral vacuum energy, the first real novelty compared to standard perturbation theory.

The chirality will be explored in more detail in Sect. 2.3.5 where I will derive the main result: The power spectrum of tensor perturbations in the Ashtekar formalism is chiral, if the Immirzi parameter γ has an imaginary part. This would lead to a non-zero TB correlator in the CMB and therefore potentially be observable.

I will finish by discussing the case of a purely real γ in Sect. 2.3.6 before concluding. Note that in the following, in general a complex value of γ will be considered, which can be split into a real and imaginary part,

$$\gamma = \gamma_R + i\gamma_I \,. \tag{2.29}$$

It will sometimes be instructive to focus on the SD/ASD connection for which $\gamma = \pm i$, or a purely imaginary γ, as these cases exhibit special behaviour. The case of a purely real γ, which renders the connection real, will not be considered until Sect. 2.3.6.

2.3.1 The Canonical Variables

To study the tensor perturbations during inflation within the Ashtekar formalism, we will consider the metric

$$ds^2 = a^2[-d\eta^2 + (\delta_{ab} + h_{ab})dx^a dx^b] \,, \tag{2.30}$$

where $a = -\frac{1}{H\eta}$ for a de Sitter background and we have omitted the TT superscript in the perturbation h_{ab}. Note that we will use the following index convention: I and μ refer to 4D internal and space-time indices, respectively, while $i, j, \ldots$ and $a, b, \ldots$ denote the corresponding 3D indices.

We need to express the perturbations in the tetrad basis to relate it to the Ashtekar variables. To zeroth order, the metric is given by (see Sect. 2.2.1) $e^{I\,(0)}_{\mu} = a\delta^I_\mu$ and the non-zero spin connection forms are $\omega^i{}_0{}^{(0)} = He^i$.

Now consider a tetrad basis of the spacetime (2.30) including perturbations, $e^I = e^{I\,(0)} + \delta e^I$. Clearly, the time component is not perturbed, so we only care about the triads e^i. A solution for the triad components that satisfies the defining relation (2.2) to first order (i.e. ignoring second order perturbations) is

$$e^i_a = a\left(\delta^i_a + \frac{1}{2}h^i_a\right) . \tag{2.31}$$

Instead of referring to the metric perturbation h_{ab}, we will simply write the perturbation in the triad as

$$e^i_a = a\delta^i_a + \delta e^i_a \,. \tag{2.32}$$

The inverse triad, which needs to satisfy Eq. (2.3) to first order is then given by

$$e^a_i = \frac{1}{a}\delta^a_i - \frac{1}{a^2}\delta e^a_i \,. \tag{2.33}$$

If we remember that δe^i_a is defined as the perturbation in the triad (2.32), we do not need to distinguish between i and a indices and can simply raise and lower them with the Kronecker delta. Although this mixes internal group and spatial indices, we can always unambiguously recover the initial perturbation δe^i_a. We will therefore refer to the perturbed triad as δe_{ij} (and simply call it the metric), and the perturbed Ashtekar connection as a_{ij}. Note that with this convention δe_{ij} will turn out to be proportional to the variable $\tilde{h}^r$ used in Sect. 1.2.5, whose mode functions v obeyed Eq. (1.62).

Like the unperturbed spin connection, its perturbation $\delta\omega^I{}_J$ must satisfy the conditions (2.15), (2.16) and (2.17) due to antisymmetry. We need to expand the torsion free equation (2.11) to first order in terms of $\delta\omega^I{}_J$, $\delta e^I{}_J$ and the unperturbed quantities $\omega^I{}_J{}^{(0)}$ and $e^{I\,(0)}_J$. For $I = 0$, we have to solve

$$\omega^0{}_i{}^{(0)} \wedge \delta e^i + \delta\omega^0{}_i \wedge e^{i\,(0)} = 0 \,, \tag{2.34}$$

where we only kept non-zero spin connection terms and used $\delta e^0 = 0$. Similarly, for $I = i$, we obtain

$$d\delta e^i + \delta\omega^i{}_0 \wedge e^{0(0)} + \delta\omega^i{}_j \wedge e^{j(0)} = 0 \,. \tag{2.35}$$

Using the rules in Sect. 2.2.1, after some algebra we find

$$\delta\omega^0_i = \frac{1}{a}\delta e'_{ij}\, dx^j \,, \tag{2.36}$$

$$\delta\omega_{ij} = -\frac{2}{a}\partial_{[i}\delta e_{j]k}\, dx^k \,, \tag{2.37}$$

where we lowered spatial indices with the Kronecker delta.

We can now define the Ashteker variables perturbed to first order, Eqs. (2.25) and (2.28). Using the background solutions for the triad and spin connection, the definition of the perturbed triad in Eq. (2.33) and noting that $\det\left(e^j_b\right) = a^3$, we obtain

$$E^a_i = a^2\delta^a_i - a\delta e^a_i \,, \tag{2.38}$$

$$A^i_a = \gamma H a \delta^i_a + \frac{a^i_a}{a} \,. \tag{2.39}$$

The classical solution for the perturbed connection a^i_a is given by the perturbed spin connections, (2.36) and (2.37):

$$a_{ij} = \epsilon_{ikl}\partial_k \delta e_{lj} + \gamma \delta e'_{ij} \,. \tag{2.40}$$

Note that this condition is only supposed to be satisfied on-shell, as initially we treat metric and connection as separate variables according to the first order formalism.

To obtain the Poisson brackets for the perturbation variables (which will be promoted to commutators when quantizing), we simply need to plug in expressions (2.38) and (2.39) into the full Poisson brackets (2.24). This results in four Poisson bracket terms of which only the last one is non-zero, which determines the Poisson bracket for fluctuations as

$$\{a^i_a(\mathbf{x}), \delta e^b_j(\mathbf{y})\} = -\frac{\gamma}{m^2_{\mathrm{Pl}}}\delta^b_a \delta^i_j \delta(\mathbf{x} - \mathbf{y}) \,. \tag{2.41}$$

2.3.2 Hamiltonian Formalism

As we know that the Holst action (2.22) is classically equivalent to the ordinary Einstein-Hilbert action (1.8), the perturbed Ashtekar variables must lead to an equation of motion for the tensor perturbations that is identical to the one you would obtain in the second order formalism. The triad satisfies $\delta e_{ij} = a h_{ij}/2$, which has the same form as the field redefinition of the tensor modes in Sect. 1.2.5, $\tilde{h}^r_{\mathbf{k}} \equiv \frac{a}{2} m_{\text{Pl}} h^r_{\mathbf{k}}$, up to a factor of m_{Pl}. It should therefore also obey the mode equation (1.62). We can obtain the equation of motion for the perturbation δe_{ij} from Hamilton's equations (derived for the full Ashtekar variables) by keeping only the first order part. Later in this section I will derive the same equations from a perturbed Hamiltonian instead.

The Hamiltonian constraint in the Ashtekar formalism for a general γ is given by [19]:

$$\mathcal{H} = \frac{m_{\text{Pl}}^2}{2} \int d^3x N E^a_i E^b_j \Big[\epsilon_{ijk}(F^k_{ab} + H^2 \epsilon_{abc} E^c_k) - 2(1+\gamma^2) K^i_{[a} K^j_{b]} \Big] . \tag{2.42}$$

Let me define the new quantities appearing in (2.42): The field strength F^i of the Ashtekar connection A^i is given by

$$F^i_{ab} = \partial_a A^i_b - \partial_b A^i_a + \epsilon^{ijk} A^j_a A^k_b , \tag{2.43}$$

K^i is the extrinsic curvature of the spatial surfaces,

$$K^i_a = \frac{A^i_a - \omega^i_a(E)}{\gamma} \tag{2.44}$$

(on shell this becomes $K^i_a \approx \omega^{0i}_a$) and $N = 1/a^2$ is the lapse density. For a SD/ASD connection, $\gamma = \pm i$, the term involving the extrinsic curvature vanishes, greatly simplifying the constraint.

We also need to take into account a Hamiltonian boundary term [46–48],

$$\mathcal{H}_{BT} = -m_{\text{Pl}}^2 \int d\Sigma_a N \epsilon_{ijk} E^a_i E^b_j A_{bk} . \tag{2.45}$$

Although the boundary term is often ignored by imposing fall-off condition at infinity [48, 49], this cannot be done in general, e.g. when using a plane wave expansion. Therefore, it will turn out to be essential to include the boundary term in order to recover the correct classical solution.

The full Hamiltonian has two other constraints [19] [as was shown in the Holst action (2.22)], the Gauss constraint

$$G_i = D_a E^a_i = \partial_a E^a_i + \epsilon_{ijk} A^j_a E^a_k \approx 0 , \tag{2.46}$$

and the vector constraint

$$V_b = E^a_i F^i_{ab} \approx 0 , \tag{2.47}$$

which is a linear combination of the Gauss and diffeomorphism constraint. Both constraints are satisfied by the background solution. It can be checked that they are also satisfied to first order using the perturbed variables (2.38) and (2.39). We will usually not be concerned with these constraints, as they do not encode the dynamics of the theory, but I will comment on their significance when perturbing the Hamiltonian to second order later.

2.3.2.1 Hamilton's Equations

To derive Hamilton's equations for the full Ashtekar variables, we need to make use the Poisson brackets in Eq. (2.24) and remember the rule $\{A, BC\} = \{A, B\}C + B\{A, C\}$. Hamilton's equations for $\gamma = \pm i$ (where the terms proportional to $(1+\gamma^2)$ in Eq. (2.42) can be ignored) take a fairly concise form:

$$A_a^{i\,\prime} = \{A_a^i, \mathcal{H}\} = \gamma N \epsilon_{ijk} E_j^b \left(F_{ab}^k + \frac{3}{2} H^2 \epsilon_{abc} E_k^c \right), \tag{2.48}$$

$$E_i^{a\,\prime} = \{E_i^a, \mathcal{H}\} = -\gamma \epsilon_{ijk} D_b (N E_j^a E_k^b) , \tag{2.49}$$

where D_a is the covariant derivative taken with the connection A^i. We can obtain evolution equations for the perturbations by plugging Eqs. (2.38) and (2.39) into (2.48) and (2.49) and expanding to first order. This gives the Hamilton equations for the perturbations,

$$a_{ij}' = 2\gamma H^2 a^2 \delta e_{ij} - \gamma \epsilon_{inm} \partial_n a_{mj} , \tag{2.50}$$

$$\delta e_{ij}' = \frac{1}{\gamma} (a_{ij} - \epsilon_{inm} \partial_n \delta e_{mj}) . \tag{2.51}$$

Hamilton's equation for δe_{ij} is the same as (2.40), i.e. it simply encodes the torsion free condition which must be satisfied on shell. Taking the derivative of (2.51), and eliminating the time derivative of the perturbed connection through (2.50), makes it possible to obtain a second order equation for δe_{ij}, independent of the connection:

$$\delta e_{ij}'' - \left(\nabla^2 + \frac{2}{\eta^2} \right) \delta e_{ij} = 0 . \tag{2.52}$$

This is the same as Eq. (1.62) in real space, proving that classically, the standard formalism of cosmological perturbation theory and the Ashtekar framework are equivalent, at least for the case $\gamma = \pm i$. Note that γ has dropped out of the equation, as it should not affect any classical results.

The Hamiltonian (2.42) of the Ashtekar formalism has been chosen such that it can be related to the ordinary Einstein-Hilbert action by a change of variables, for any choice of γ. Therefore we know that Eq. (2.52) needs to hold in the general case as well. This will help us in deriving Hamilton's equations for the perturbations. For a general γ, Hamilton's equations, derived for the full Ashtekar variables, are a lot more complicated than in the SD/ASD case. Taking the Poisson brackets with the

Hamiltonian (2.42), we obtain the same expression as in Eqs. (2.48) and (2.49), plus additional terms proportional to $(1+\gamma^2)$:

$$A_a^{i\,\prime} = \gamma N \epsilon_{ijk} E_j^b \left(F_{ab}^k + \frac{3}{2} H^2 \epsilon_{abc} E_k^c \right) - \gamma(1+\gamma^2) N E_j^b (K_b^j K_a^i - K_a^j K_b^i)$$

$$-m_{\rm Pl}^2 (1+\gamma^2) \int d^3y N E_j^b E_k^c \{A_a^i(x), \omega_{[b}^j \omega_{c]}^k\} \qquad (2.53)$$

$$E_i^{a\,\prime} = -\gamma \epsilon_{ijk} D_b (N E_j^a E_k^b) + (1+\gamma^2) N (E_i^a E_j^b - E_j^a E_i^b) K_b^j \,. \qquad (2.54)$$

The Poisson bracket $\{A, \omega(E)\}$ is a very long and messy expression, so the last term of Eq. (2.53) is left unexpanded. As for the case $\gamma = \pm i$, we can obtain the evolution equations for the perturbations by substituting the definition of the Ashtekar variables into (2.53) and (2.54) and expanding to first order.

In the case of the triad, this yields the same expression as before, Eq. (2.51). We would like to avoid having to work out Hamilton's equation for a_{ij} explicitly as it would involve having to compute the unexpanded Poisson bracket in (2.53). As we know that e_{ij} needs to satisfy the equation of motion (2.52), we can actually avoid doing the explicit calculation and simply use Eqs. (2.51) and (2.52) to deduce Hamilton's equation for the connection. It should contain the terms on the RHS of Eq. (2.50) plus additional terms proportional to $(1+\gamma^2)$, such that it reduces to the old expression for $\gamma = \pm i$. Carrying out these manipulations, we finally obtain Hamilton's equations for the perturbations for a general value of γ:

$$a'_{ij} = 2\gamma H^2 a^2 \delta e_{ij} - \gamma \epsilon_{inm} \partial_n a_{mj} + \frac{1+\gamma^2}{\gamma} \epsilon_{inm} \partial_n (a_{mj} - \epsilon_{mkl} \partial_k \delta e_{lj}) \,, \qquad (2.55)$$

$$\delta e'_{ij} = \frac{1}{\gamma} (a_{ij} - \epsilon_{inm} \partial_n \delta e_{mj}) \,. \qquad (2.56)$$

At first glance, it might seem odd that these expressions yield the same equation of motion for the triad [Eq. (2.52)] as in the case $\gamma = \pm i$, considering the connection equation has acquired an additional term in $1+\gamma^2$ compared to Eq. (2.50). However, this is necessary as terms proportional to $1+\gamma^2$ do actually appear in the derivation of the result for the $\gamma = \pm i$ case, where they can be set to zero. These terms must be present in the case of general γ.

2.3.2.2 Second Order Hamiltonian

We have found the equations of motion for the perturbations by perturbing the full Hamilton equations. However, to be able to quantize the theory, we need to identify the perturbed Hamiltonian. This exercise is not trivial; as we will see in the following, a fair number of subtleties need to be taken into account before arriving at the correct result.

The perturbed Hamiltonian should contain tensor perturbations and encode the dynamics of gravitons. Therefore, we know that the constraint $\mathcal{H} \approx 0$, which demonstrates the lack of dynamics, cannot apply to the perturbative Hamiltonian which we would like to quantize. Let us think about the Hamiltonian to different orders in the perturbative expansion.

The first order Hamiltonian is trivially zero (once the other constraints are used). The second order Hamiltonian, on the other hand, includes two terms,

$$ {}^{2}\mathcal{H} = {}^{2}_{1}\mathcal{H} + {}^{2}_{2}\mathcal{H}\,, \tag{2.57}$$

where ${}^{2}_{1}\mathcal{H}$ contains products of first order perturbations, and ${}^{2}_{2}\mathcal{H}$ is linear in second order perturbations in the triad and connection. Only the sum of these terms vanishes on shell, ${}^{2}\mathcal{H} \approx 0$. We can therefore identify the first term, ${}^{2}_{1}\mathcal{H}$, with the dynamical Hamiltonian to second order, while the second term ${}^{2}_{2}\mathcal{H}$ simply encodes the backreaction or compensation due to the non-linearity of the gravitational field, which ensures that the Hamiltonian constraint is satisfied. Therefore, we will need to calculate ${}^{2}_{1}\mathcal{H}$ to understand graviton dynamics.

Let me also stress that in the Ashtekar formulation, off-shell, *the Hamiltonian is not real*, due to the presence of the complex Immirzi parameter γ. Of course, imposing the constraints, the Hamiltonian becomes weakly zero and is therefore manifestly real. However, as the constraint does not apply to the dynamical second order Hamiltonian, ${}^{2}_{1}\mathcal{H}$ is indeed complex. The complexity of ${}^{2}_{1}\mathcal{H}$ will have an effect on perturbation theory, and the novelties I will describe can be traced back to this fact. Even though a complex Hamiltonian might seem strange, the quantum theory we set up later (Sect. 2.3.4) will still be well defined. All classical results can be recovered and the quantum Hamiltonian will turn out to be hermitian after fixing the inner product.

Before proceeding, note that the other constraints are also not zero when considering only the second order part that is quadratic in first order perturbations. Specifically, for the Gauss constraint we get

$$ {}^{2}_{1}G_i = -\epsilon_{ijk} a^j_a \delta e^a_k \neq 0\,. \tag{2.58}$$

When deriving (2.42) from the usual ADM action, the Gauss constraint and the torsion free condition are used [19]. Therefore, non-zero terms proportional to ${}^{2}_{1}G_i$ and ${}^{2}_{1}T^a$ will appear in the expression for ${}^{2}_{1}\mathcal{H}$. However, it can be checked that these additional terms result in a full divergence and can therefore be ignored.

By expanding the Hamiltonian (2.42) to second order we obtain:

$$ \begin{aligned} {}^{2}_{1}\mathcal{H} = \frac{m_{\rm Pl}^2}{2}\int d^3x \bigg\{ & \frac{1}{\gamma^2} a_{ij}a_{ij} + 2\epsilon_{ijk}\delta e_{li}\partial_j a_{kl} - 2H^2a^2\delta e_{ij}\delta e_{ij} \\ & + \frac{2}{\gamma} Ha\delta e_{ij}a_{ij} - 2\frac{1+\gamma^2}{\gamma} Ha\delta e_{ij}\epsilon_{ikl}(\partial_k\delta e_{lj}) \\ & - \frac{1+\gamma^2}{\gamma^2}\Big[\epsilon_{ikl}(\partial_k\delta e_{lj})a_{ij} + \epsilon_{ikl}a_{ij}(\partial_k\delta e_{lj}) - \epsilon_{ikl}\epsilon_{jmn}(\partial_k\delta e_{lj})(\partial_m\delta e_{ni})\Big]\bigg\}\,, \end{aligned} \tag{2.59}$$

where we kept the ordering as it appeared in the calculation, as it will affect the quantization. Only the first four terms survive for $\gamma = \pm i$. This expression is not the correct perturbative Hamiltonian, however: it does not reduce to the Hamiltonian (1.59) obtained for tensor perturbations in the second order formalism on shell, i.e. when using the torsion free condition (2.40). This is due to two reasons.

First, we have not yet included the boundary term (2.45) at the same order and level in perturbation theory (second order terms quadratic in first order variables). It is given by

$$ {}^2_1\mathcal{H}_{BT} = m_{\rm Pl}^2 \int d\Sigma_i \epsilon_{ijk} \delta e_{lj} a_{lk} \,. \tag{2.60}$$

To make this into a volume instead of a surface integral, we use Stokes' theorem [50] which introduces a divergence,

$$ {}^2_1\mathcal{H}_{BT} = \frac{m_{\rm Pl}^2}{2} \int d^3x \; 2\epsilon_{ijk} \partial_i (\delta e_{lj} a_{lk}) \,, \tag{2.61}$$

where we introduced factors of two to obtain the same pre-factor as in (2.59). The derivative term can be split into two contributions, one of which cancels the second term of Eq. (2.59) and the other is $-2\epsilon_{ijk}(\partial_j \delta e_{li})a_{kl}$.

The second issue is more subtle and related to the terms proportional to H in (2.59). There should not be any terms linear in the Hubble rate, as we want to rederive the Hamiltonian for ordinary tensor perturbations, Eq. (1.59), where the only explicitly time dependent term is a''/a, which in de Sitter is given by $2/\eta^2 = 2a^2H^2$ and is therefore quadratic in H.

To understand what has gone wrong, recall the perturbed expression for the triad and connection:

$$ A^i_a = \gamma H a \delta^i_a + \frac{a^i_a}{a} \,, \tag{2.62}$$

$$ E^a_i = a^2 \delta^a_i - a \delta e^a_i \,. \tag{2.63}$$

Instead of thinking of this as a zero order part plus a perturbation, you can also regard it as a canonical transformation [51]: we have replaced variables (A^i_a, E^b_j) with variables $(a^i_a, \delta e^b_j)$, which have the same symplectic structure as the original variables (the fact that the Poisson brackets (2.41) have a minus sign compared to (2.24) is related to the fact that we defined the perturbation δe^b_j in the densitized triad, not its inverse, initially. We could also absorb the minus sign into a field redefinition of the triad perturbation). Such a transformation can always be performed for canonical systems, regardless of whether the new variables are small perturbations. In this viewpoint, instead of "freezing" the background and considering spacetime perturbations around it, we regard the perturbed variables as equivalent to the full Ashtekar variables.

If the canonical transformation is explicitly time dependent (which it is as a is a function of time), the Hamiltonian in terms of the new variables, denoted by $\mathcal{K}$, is related to the old Hamiltonian by a generating function F [51]:

$$\mathcal{K} = \mathcal{H} + \frac{\partial F}{\partial \eta} . \tag{2.64}$$

To obtain the correct Hamiltonian, in principal we therefore need to compute the generating function. However, again it is possible to "cheat" slightly by using consistency arguments instead of performing explicit calculations. We know that it should be possible to derive Hamilton's equations for the perturbations by taking the Poisson brackets (2.41) with the (correct) perturbed Hamiltonian to second order. By demanding consistency with Eqs. (2.55) and (2.56), which were obtained from perturbing the full Hamilton's equations, we find that the appropriate generating function must be

$$\frac{\partial F}{\partial \eta} = -\frac{m_{\rm Pl}^2}{\gamma} \int d^3x H a \delta e_{ij} \left[a_{ij} - (1+\gamma^2)\epsilon_{ikl}\partial_k \delta e_{lj} \right] .$$

Adding this term to the Hamiltonian in (2.59) eliminates the second line, i.e. the terms proportional to H. The final expression, taking the boundary term (2.61) into account, is therefore:

$$\begin{aligned} \mathcal{H}_{eff} = \frac{m_{\rm Pl}^2}{2} \int d^3x \Bigg[& \frac{1}{\gamma^2} a_{ij} a_{ij} - 2H^2 a^2 \delta e_{ij} \delta e_{ij} + \left(1 - \frac{1}{\gamma^2}\right) \epsilon_{ikl} (\partial_k \delta e_{lj}) a_{ij} \\ & - \left(1 + \frac{1}{\gamma^2}\right) \epsilon_{ikl} a_{ij} (\partial_k \delta e_{lj}) + \left(1 + \frac{1}{\gamma^2}\right) \epsilon_{ikl} \epsilon_{jmn} (\partial_k \delta e_{lj})(\partial_m \delta e_{ni}) \Bigg] . \end{aligned} \tag{2.65}$$

This corresponds to the effective perturbative Hamiltonian, which can be used to quantize the theory in terms of graviton states.

By using the on-shell condition (2.40), we can derive the Hamiltonian in terms of the triad only, remembering $2a^2 H^2 = a''/a$:

$$\mathcal{H}_{eff}\big|_{\rm on-shell} = \frac{m_{\rm Pl}^2}{2} \int d^3x \left[\delta e'_{ij} \delta e'_{ij} + (\partial_k \delta e_{ij})^2 - \frac{a''}{a} \delta e_{ij} \delta e_{ij} \right] . \tag{2.66}$$

After identifying the two physical polarizations of the triad by using appropriate mode expansions in the next section, it will be clear that this is exactly the same as expression as (1.59), the second order Hamiltonian for tensor modes derived in the second order formalism.

2.3.3 Fourier Space Expansion

To be able to quantize the theory, we need to expand the perturbed variables in terms of Fourier modes. However, we need to be careful that we perform this expansion correctly, by taking into account two separate, but related points.

Firstly, note that in the Ashtekar formalism, the connection is initially complex and we are not enforcing any reality conditions before quantizing. Therefore, we must have graviton and anti-graviton modes in the expansion. This means that the negative and positive frequencies in the field expansion are initially independent (so compared to Eq. (1.45), we should have a different operator $b_{\mathbf{k}}$ associated with the second term). Secondly, we will make the common field theory choice stipulating that the spatial vector $\mathbf{k}$ points in the direction of propagation for both positive and negative frequency terms. The reality conditions will then identify gravitons and anti-gravitons moving in the same direction, not in opposite directions.

This choice not always been made in previous literature on the subject, where non-physical couplings between $\mathbf{k}$ and $-\mathbf{k}$ modes appeared in the physical Hamiltonian inside the horizon [49, 52]. These should only be present outside the horizon, where they represent the production of particle pairs by the gravitational field (with the particles in each pair moving in opposite directions) [53].

We therefore make the following Fourier expansion:

$$\delta e_{ij} = \int \frac{d^3k}{(2\pi)^{\frac{3}{2}}} \sum_r \epsilon^r_{ij}(\mathbf{k})\tilde{e}_{r+}(\mathbf{k},\eta)e^{i\mathbf{k}\cdot\mathbf{x}} + \epsilon^{r\star}_{ij}(\mathbf{k})\tilde{e}^{\dagger}_{r-}(\mathbf{k},\eta)e^{-i\mathbf{k}\cdot\mathbf{x}},$$

$$a_{ij} = \int \frac{d^3k}{(2\pi)^{\frac{3}{2}}} \sum_r \epsilon^r_{ij}(\mathbf{k})\tilde{a}_{r+}(\mathbf{k},\eta)e^{i\mathbf{k}\cdot\mathbf{x}} + \epsilon^{r\star}_{ij}(\mathbf{k})\tilde{a}^{\dagger}_{r-}(\mathbf{k},\eta)e^{-i\mathbf{k}\cdot\mathbf{x}}, \quad (2.67)$$

where $\tilde{e}_{rp}(\mathbf{k},\eta) = e_{rp}(\mathbf{k})\Psi_e(k,\eta)$ and $\tilde{a}_{rp}(\mathbf{k},\eta) = a_{rp}(\mathbf{k})\Psi^{rp}_a(k,\eta)$, and ϵ^r_{ij} are polarization tensors. In a frame where the direction $i = 1$ is aligned with $\mathbf{k}$, they are given by [54]:

$$\epsilon^{(r)}_{ij} = \frac{1}{\sqrt{2}}\begin{pmatrix} 0 & 0 & 0 \\ 0 & 1 & \pm i \\ 0 & \pm i & -1 \end{pmatrix}. \quad (2.68)$$

Equation (2.67) has the same form as the mode expansion for tensor perturbations (1.57) performed in Sect. 1.2.5, but now with an additional negative frequency term which is independent of the first, as required. The amplitudes $a_{rp}(\mathbf{k})$ and $e_{rp}(\mathbf{k})$ have two indices (in contrast with some of the previous literature [49, 52]): $r = \pm 1$ for right (R) and left (L) helicities, and p for graviton ($p = 1$) and anti-graviton ($p = -1$) modes (which were not present in Eq. (1.57), where the tensor perturbations were manifestly real).

We can assume that the amplitudes a_{rp} and e_{rp}, which will correspond to annihilation operators upon quantization, are equal, and the differences can be absorbed into the mode functions Ψ_e and Ψ_a. Imposing the on-shell condition we will find that while Ψ_e is independent of helicity and graviton states, the mode functions for the connection, $\Psi_a(k, \eta)$, must carry an r, p dependence.

2.3.3.1 Mode Functions

As we have seen that the Ashtekar formalism is equivalent to the second order formalism of Sect. 1.2.5, we know that the mode functions of the triad will satisfy the equation of motion (1.62)

$$\Psi_e'' + \left(k^2 - \frac{2}{\eta^2}\right)\Psi_e = 0, \tag{2.69}$$

where $'$ denotes differentiation with respect to conformal time. This has the Bunch-Davies solution given in Eq. (1.62),

$$\Psi_e = \frac{e^{-ik\eta}}{2\sqrt{k}}\left(1 - \frac{i}{k\eta}\right). \tag{2.70}$$

The boundary condition in the far past, $|k\eta| \gg 1$, is

$$\Psi(k, \eta) \sim e^{-ik\eta}. \tag{2.71}$$

This shows that **k** can be regarded as the direction of propagation of the wave as the exponentials in which **k** appears can be written in four-vector form as $e^{-ik\eta}e^{i\mathbf{k}\cdot\mathbf{x}} = e^{ik_\mu x^\mu}$, $k_\mu k^\mu = 0$.

Let us find an expression for the mode functions of the connection on-shell. We need to plug the Fourier space expansion (2.67) into the classical solution of the connection derived from the torsion free condition, Eq. (2.40). Making use of the identity

$$\epsilon_{inl} k_n \epsilon_{lj}^{(r)} = -irk\epsilon_{ij}^{(r)}, \tag{2.72}$$

we find

$$\Psi_a^{rp} = (\gamma_R + p\gamma_I)\Psi_e' + rk\Psi_e. \tag{2.73}$$

This expression can be simplified inside the horizon ($k|\eta| \gg 1$), when the boundary condition (2.71) holds:

$$\Psi_a^{rp} = \Psi_e k\,(r - i\gamma_R + p\gamma_I)\ , \tag{2.74}$$

There is only a dependence on p if γ has an imaginary part and for a purely real γ, Ψ_a^r would be the same for gravitons and anti-gravitons. This is to be expected, as

Table 2.1 Relationship between helicity and duality states

	$r = +$ [R]	$r = -$ [L]
$p = +$ $[G]$	SD	ASD
$p = -$ $[\overline{G}]$	ASD	SD

for a manifestly real theory we would not have needed to expand in terms of two different operators a_{r+} and a_{r-}, but just a single a_r.

Before carrying on with the quantization of the perturbations, let us briefly investigate the relationship between the helicity states, labelled by r, and the duality states, defined by $\gamma = \pm i$. In this case, Eq. (2.74) becomes

$$\Psi_a^{rp} = (r - ip\gamma)k\Psi_e \ . \tag{2.75}$$

For an SD connection, $i\gamma = -1$, and the quantity in brackets is simply $(r + p)$. This is clearly zero if r and p have different signs. Therefore, the only components of the connection that survive in the self dual case are the right handed ($r = 1$) positive frequency of the graviton ($p = 1$) and the left handed ($r = -1$) negative frequency of the anti-graviton ($p = -1$). The ASD connection has $i\gamma = +1$ and therefore contains the remaining degrees of freedom, right handed anti-graviton and left-handed graviton. The split of the states into SD and ASD parts is summarized in Table 2.1.

This analysis shows that helicity modes and duality modes do not align, i.e. the SD connection carries both right and left-handed helicity states and similarly for the ASD connection. This point has been highlighted in [55], but it requires performing the correct Fourier space expansion including graviton and anti-gravitons states and was therefore missed in [49, 52].

2.3.3.2 Reality Conditions

When we set up the Hilbert space of quantum states in Sect. 2.3.4, we will need to impose reality conditions to relate graviton and anti-graviton states (and their Hermitian conjugates), which will enable us to obtain the physical degrees of freedom. The reality conditions will eventually be used to fix the inner product, but it is instructive to obtain the corresponding conditions on the operators.

The metric is real, $\delta e_{ij} = \overline{\delta e}_{ij}$. Imposing this on the Fourier expansion, we find

$$e_{r+}(\mathbf{k}) = e_{r-}(\mathbf{k}) \ . \tag{2.76}$$

Therefore, graviton and anti-graviton are identified for each polarization and each mode $\mathbf{k}$. This is a good check that the expansion in Eq. (2.67) is physically sensible, as we do not get relations between different polarizations or wavevectors $\mathbf{k}$ and $-\mathbf{k}$.

On-shell, the triad therefore only needs one set of creation and annihilation operators in its Fourier expansion.

For the connection, the torsion free condition and the reality condition are linked: Although the connection can be complex, it must satisfy the torsion-free condition, which will ensure that the metric is real. From the defining expression for the Ashtekar connection, Eq. (2.28), we know that

$$\Re A^i = \omega^i + \gamma_R \omega^{0i} , \tag{2.77}$$
$$\Im A^i = \gamma_I \omega^{0i} . \tag{2.78}$$

There are two reality conditions for the connection, but we only need to impose one as a constraint, as the dynamical evolution (described by Hamilton's equations) will make sure that the second condition is satisfied. Let us see what this implies for the perturbations a_{ij}. Using the solutions for the perturbed spin connection components, Eqs. (2.36) and (2.37), we obtain

$$a_{ij} + \bar{a}_{ij} = 2a\left(\delta\omega_{ij} + \gamma_R \delta\omega^0_{ij}\right) = 2\epsilon_{ikl}\partial_k \delta e_{lj} + 2\gamma_R \delta e'_{ij} , \tag{2.79}$$
$$a_{ij} - \bar{a}_{ij} = 2ai\gamma_I \delta\omega^0_{ij} = 2i\gamma_I \delta e'_{ij} . \tag{2.80}$$

Using the expansion (2.67), in Fourier space this becomes

$$\tilde{a}_{r+}(\mathbf{k}, \eta) + \tilde{a}_{r-}(\mathbf{k}, \eta) = 2rk\tilde{e}_{r+}(\mathbf{k}, \eta) + 2\gamma_R \tilde{e}'_{r+}(\mathbf{k}, \eta) , \tag{2.81}$$
$$\tilde{a}_{r+}(\mathbf{k}, \eta) - \tilde{a}_{r-}(\mathbf{k}, \eta) = 2i\gamma_I \tilde{e}'_{r+}(\mathbf{k}, \eta) , \tag{2.82}$$

where $\tilde{a}_{rp} = a_{rp}\Psi_a^{rp}$ and $\tilde{e}_{rp} = e_{rp}\Psi_e$. The reality condition for the connection we want to impose as a constraint should be non-dynamical, so let us eliminate the time derivative of the metric by combining Eqs. (2.81) and (2.82):

$$i\gamma^* \tilde{a}_{r+}(\mathbf{k}, \eta) - i\gamma \tilde{a}_{r-}(\mathbf{k}, \eta) = 2rk\gamma_I \tilde{e}_{r+}(\mathbf{k}, \eta) . \tag{2.83}$$

Its Hermitian conjugate is:

$$-i\gamma \tilde{a}^\dagger_{r+}(\mathbf{k}, \eta) + i\gamma^* \tilde{a}^\dagger_{r-}(\mathbf{k}, \eta) = 2rk\gamma_I \tilde{e}^\dagger_{r-}(\mathbf{k}, \eta) , \tag{2.84}$$

where we have used Eq. (2.76) to turn $p = 1$ into $p = -1$ on the RHS. This shows that for each r and $\mathbf{k}$ there are two independent conditions upon the four operators $a_{rp}(\mathbf{k})$ and $e_{rp}(\mathbf{k})$. We will use them later when we define the inner product.

On shell, we can use the full torsion-free conditions, Eq. (2.73), which can be written as a weak identity on the operators:

$$\tilde{a}_{r-}(\mathbf{k}, \eta) \approx rk\tilde{e}_r + \gamma^* \tilde{e}'_r \to \tilde{e}_r(r - i\gamma^*)k , \tag{2.85}$$
$$\tilde{a}_{r+}(\mathbf{k}, \eta) \approx rk\tilde{e}_r + \gamma \tilde{e}'_r \to \tilde{e}_r(r - i\gamma)k , \tag{2.86}$$

where the latter expression is valid in the limit $k|\eta| \gg 1$, c.f. Eq. (2.74). These identities will be useful later when deriving the graviton operators for this theory, as they will show that one of the graviton modes is unphysical.

2.3.3.3 Commutation Relations

Before we can set up a quantum theory in terms of graviton operators we need to define the commutation relations for the modes. To do this, we first promote the Poisson brackets (2.24) and (2.41) of the connection and metric in position space to commutators:

$$\left[A^i_a(\mathbf{x}), E^b_j(\mathbf{y})\right] = i\frac{\gamma}{m^2_{\text{Pl}}}\delta^b_a\delta^i_j\delta(\mathbf{x}-\mathbf{y})\,, \tag{2.87}$$

$$\left[a^i_a(\mathbf{x}), \delta e^b_j(\mathbf{y})\right] = -i\frac{\gamma}{m^2_{\text{Pl}}}\delta^b_a\delta^i_j\delta(\mathbf{x}-\mathbf{y})\,. \tag{2.88}$$

Note that these commutators have been derived from the fundamental Poisson brackets of the Ashtekar variables and hence have not been gauge fixed yet, i.e. the TT projection has not been carried out and we therefore have not identified the two physical polarizations of tensor perturbations. The Fourier expansion (2.67), on the other hand, assumed by construction that there are only two helicity states $r = \pm 1$. It was shown in [56] that the appropriate form of the commutator (2.88), taking care of the gauge fixing, is

$$[a_{ij}(\mathbf{x}), \delta e_{kl}(\mathbf{y})] = -i\frac{\gamma}{m^2_{\text{Pl}}}P_{ijkl}(\mathbf{x}-\mathbf{y})\,, \tag{2.89}$$

where the delta function is replaced by a function $P_{ijkl}(\mathbf{x})$ which takes care of the TT projection and is given by

$$P_{ijkl}(\mathbf{x}) = \int \frac{d^3k}{(2\pi)^3}\sum_r \epsilon^r_{ij}(\mathbf{k})\epsilon^{r\star}_{kl}(\mathbf{k})e^{i\mathbf{k}\cdot\mathbf{x}}\,. \tag{2.90}$$

To obtain the equivalent of Eq. (2.89) for modes, let us first consider the unprojected commutator (2.88) again. Dropping the indices, we can split the metric and connection into separate positive and negative frequency parts, $\delta e = \delta e^+ + \delta e^-$, $a = a^+ + a^-$, which are given by

$$\delta e^+(\mathbf{x},\eta) = \int \frac{d^3k}{(2\pi)^{\frac{3}{2}}}e^+(\mathbf{k},\eta)e^{i\mathbf{k}\cdot\mathbf{x}}\,, \tag{2.91}$$

$$\delta e^-(\mathbf{x},\eta) = \int \frac{d^3k}{(2\pi)^{\frac{3}{2}}}e^{-\dagger}(\mathbf{k},\eta)e^{-i\mathbf{k}\cdot\mathbf{x}}\,, \tag{2.92}$$

and similarly for a.

Therefore there are four terms in the commutator and, as is standard in QFT [6], the only non-vanishing equal-time commutators must be given by positive and negative frequency parts,

$$[a^{+}(\mathbf{x}), \delta e^{-}(\mathbf{y})] = [a^{-}(\mathbf{x}), \delta e^{+}(\mathbf{y})] = -i\frac{\gamma}{2m_{\rm Pl}^{2}}\delta(\mathbf{x}-\mathbf{y}) \; . \tag{2.93}$$

For the modes, this implies

$$[a^{+}(\mathbf{k}), e^{-\dagger}(\mathbf{k}')] = [a^{-}(\mathbf{k}), e^{+\dagger}(\mathbf{k}')] = -i\frac{(\gamma_R + p i \gamma_I)}{2m_{\rm Pl}^{2}}\delta(\mathbf{k}-\mathbf{k}') \; , \tag{2.94}$$

Taking expression (2.89) for the TT projected position space commutators into account, we see that the operators we have defined in the Fourier expansion (2.67) have commutation relation

$$[\tilde{a}_{rp}(\mathbf{k}), \tilde{e}_{sq}^{\dagger}(\mathbf{k}')] = -i\frac{(\gamma_R + p i \gamma_I)}{2m_{\rm Pl}^{2}}\delta_{rs}\delta_{p\bar{q}}\delta(\mathbf{k}-\mathbf{k}') \; , \tag{2.95}$$

where $\overline{q} = -q$.

The dependence on $\delta_{p\bar{q}}$ shows that we only get non-vanishing commutators when considering the positive frequency of one variable and the negative frequency of the other. As before, when we considered the mode functions of the connection, there is no p dependence if $\gamma_I = 0$, as for a real field there is no distinction between gravitons and anti-gravitons.

2.3.4 *Quantum Hamiltonian*

We now have all the ingredients to set up the Hamiltonian in Fourier space which will be the starting point for the quantum theory. We want to express it in the standard form where it just reduces to a creation times an annihilation operator, counting the number of states, c.f. Eq. (1.44). In our case, these states will be graviton states and the operators will create and annihilate gravitons. As we have not imposed the torsion-free condition yet, the graviton operators will be linear combinations of the metric and connection, and only reduce to metric variables on-shell. Due to the complexity of the Hamiltonian, this exercise is non-trivial. We will find twice as many particle states as expected as well as unphysical particle production terms. However, once the correct inner product has been identified, we will reproduce the expected form of the Hamiltonian.

Note that from now on we will consider the inside horizon limit $k\eta \gg 1$ for which terms in H can be neglected, as we are not interested in the behaviour of tensor perturbations outside the horizon where they freeze out.

Inserting the expansion (2.67) into (2.65) and making use of the relations

$$\epsilon^r_{ij}(\mathbf{k})\epsilon^{s\star}_{ij}(\mathbf{k}) = 2\delta^{rs}\,, \quad \epsilon^r_{ij}(-\mathbf{k}) = \epsilon^{r\star}_{ij}(\mathbf{k})\,, \tag{2.96}$$

we obtain a lengthy expression for the Fourier space Hamiltonian:

$$\begin{aligned}\mathcal{H}_{eff} = m_{\rm Pl}^2 \int d^3k \sum_r \frac{1}{\gamma^2}\Big\{ \\ &\Big[k^2\left(\gamma^2+1\right)\tilde{e}_{r+}(\mathbf{k}) - kr\left(\gamma^2+1\right)\tilde{a}_{r+}(\mathbf{k})\Big]\tilde{e}_{r+}(-\mathbf{k}) \\ &+\Big[k^2\left(\gamma^2+1\right)\tilde{e}_{r+}(\mathbf{k}) - kr\left(\gamma^2+1\right)\tilde{a}_{r+}(\mathbf{k})\Big]\tilde{e}^\dagger_{r-}(\mathbf{k}) \\ &+\Big[k^2\left(\gamma^2+1\right)\tilde{e}^\dagger_{r-}(\mathbf{k}) - kr\left(\gamma^2+1\right)\tilde{a}^\dagger_{r-}(\mathbf{k})\Big]\tilde{e}_{r+}(\mathbf{k}) \\ &+\Big[k^2\left(\gamma^2+1\right)\tilde{e}^\dagger_{r-}(\mathbf{k}) - kr\left(\gamma^2+1\right)\tilde{a}^\dagger_{r-}(\mathbf{k})\Big]\tilde{e}^\dagger_{r-}(-\mathbf{k}) \\ &+\Big[kr\left(\gamma^2-1\right)\tilde{e}_{r+}(\mathbf{k}) + \tilde{a}_{r+}(\mathbf{k})\Big]\tilde{a}_{r+}(-\mathbf{k}) \\ &+\Big[kr\left(\gamma^2-1\right)\tilde{e}_{r+}(\mathbf{k}) + \tilde{a}_{r+}(\mathbf{k})\Big]\tilde{a}^\dagger_{r-}(\mathbf{k}) \\ &+\Big[kr\left(\gamma^2-1\right)\tilde{e}^\dagger_{r-}(\mathbf{k}) + \tilde{a}^\dagger_{r-}(\mathbf{k})\Big]\tilde{a}_{r+}(\mathbf{k}) \\ &+\Big[kr\left(\gamma^2-1\right)\tilde{e}^\dagger_{r-}(\mathbf{k}) + \tilde{a}^\dagger_{r-}(\mathbf{k})\Big]\tilde{a}^\dagger_{r-}(-\mathbf{k})\Big\}\,.\end{aligned} \tag{2.97}$$

2.3.4.1 Hamiltonian for $\gamma = \pm i$

Before trying to make sense of this monstrosity, it is instructive to study the case of a SD/ASD connection for which $\gamma^2 = -1$. In this case, Eq. (2.97) reduces to a much more tractable form:

$$\begin{aligned}\mathcal{H}_{eff} &= m_{\rm Pl}^2 \int d^3k \sum_r g_{r-}(\mathbf{k})g_{r+}(-\mathbf{k}) + g_{r-}(\mathbf{k})g^\dagger_{r-}(\mathbf{k}) \\ &\quad + g^\dagger_{r+}(\mathbf{k})g_{r+}(\mathbf{k}) + g^\dagger_{r+}(\mathbf{k})g^\dagger_{r-}(-\mathbf{k})\,,\end{aligned} \tag{2.98}$$

where

$$g_{r+}(\mathbf{k}) = \tilde{a}_{r+}(\mathbf{k})\,, \tag{2.99}$$
$$g^\dagger_{r+}(\mathbf{k}) = -\tilde{a}^\dagger_{r-}(\mathbf{k}) + 2kr\tilde{e}^\dagger_{r-}(\mathbf{k})\,, \tag{2.100}$$
$$g_{r-}(\mathbf{k}) = -\tilde{a}_{r+}(\mathbf{k}) + 2kr\tilde{e}_{r+}(\mathbf{k})\,, \tag{2.101}$$
$$g^\dagger_{r-}(\mathbf{k}) = \tilde{a}^\dagger_{r-}(\mathbf{k})\,, \tag{2.102}$$

which can be identified as the the graviton ($p = 1$) and anti-graviton ($p = -1$) creation and annihilation operators $g^{\dagger}_{rp}$, g_{rp}. Note that the creation and annihilation operators for each index r, p are only hermitian conjugates of each after the reality conditions (2.83) and (2.84) have been imposed.

Their commutation relations can be derived from Eq. (2.95):

$$[g_{rp}(\mathbf{k}), g^{\dagger}_{sq}(\mathbf{k}')] = -\frac{i\gamma}{m^2_{\rm Pl}}(pr)k\delta_{rs}\delta_{pq}\delta(\mathbf{k}-\mathbf{k}') \, . \tag{2.103}$$

The Hamiltonian (2.98) has some unusual features. Firstly, for each $\mathbf{k}$ we find four independent modes ($r = \pm 1$ and $p = \pm\, 1$), instead of two as would be expected for tensor perturbations. Half of these states have negative energy (those with $i\gamma = pr$, which leads to a minus sign in the commutator instead of the usual plus sign). For example, for the SD connection $\gamma = i$ the left "graviton" ($r = -1$ and $p = 1$) and the right "anti-graviton" ($r = 1$ and $p = -1$) carry negative energy. Secondly, there are unphysical production terms in the Hamiltonian (2.98) which couple $\mathbf{k}$ and $-\mathbf{k}$ modes. These pump terms represent pair production [53], and should not be present in the subhorizon limit $k|\eta| \gg 1$ where spacetime is approximately flat.

Both of these pathological features are not present for classical solutions, as they vanish on-shell when imposing the conditions (2.85) and (2.86). For example, for $\gamma = i$, the on-shell conditions imply $a_{R-} \approx 0$ and $a_{L+} \approx 0$. When also imposing the reality conditions such that we can consider the creation and annihilation operators as hermitian conjugates of one another, $g^{\dagger}_{rp} = (g_{rp})^{\dagger}$, we find that two of the operators are eliminated. Only $g^{\dagger}_{R+}$, $g^{\dagger}_{L-}$, which create positive energy states, are non-zero. Thus, the negative energy modes do not exist classically and you can check that the pump terms also vanish.

As mentioned previously, quantum mechanically we do not want to treat the reality conditions as operator conditions but impose them on the inner product, which should also remove the unphysical states from the Hilbert space. We will use a holomorphic representation where we consider the states as analytic functions over the complex domain as introduced by Bargmann [57].

As mentioned above, the reality conditions simply ensure that $g^{\dagger}_{rp}$ is indeed the hermitian conjugate of g_{rp}. This condition is sufficient to fix the inner product [49, 58, 59]. A holomorphic representation for wavefunctions $\Phi = \langle z|\Phi\rangle$ is defined as one which diagonalises $g^{\dagger}_{rp}$ [57]:

$$\langle z|g^{\dagger}_{rp}|\Phi\rangle = z_{rp}\langle z|\Phi\rangle \, , \tag{2.104}$$

where $z_{rp}(\mathbf{k})$ are complex eigenvalues. Similarly to the case of deriving the action of the momentum operator on states when working in the usual position space representation, we can derive the action of g_{rp} from the commutator (2.103):

$$\langle z|g_{rp}|\Phi\rangle = -i\frac{\gamma}{m^2_{\rm Pl}}(pr)k\frac{\partial}{\partial z_{rp}}\langle z|\Phi\rangle \, . \tag{2.105}$$

We want to define an inner product in this representation. The decomposition of the unity operator for the complex eigenvectors $|z\rangle$ is given by [59]

$$\mathbb{1} = \int dz d\bar{z} e^{\mu(z,\bar{z})} , \tag{2.106}$$

where $e^{\mu(z,\bar{z})}$ is a positive integration measure (for the normal position representation with eigenstates $|x\rangle$, it is just equal to 1). The inner product can then be written as

$$\langle \Phi_1 | \Phi_2 \rangle = \int dz d\bar{z} e^{\mu(z,\bar{z})} \bar{\Phi}_1(\bar{z}) \Phi_2(z) . \tag{2.107}$$

The defining condition of the hermitian conjugate of an operator is $\langle \Phi_1 | g^{\dagger}_{rp} | \Phi_2 \rangle = \overline{\langle \Phi_2 | g_{rp} | \Phi_1 \rangle}$, which can be used to derive an expression for the measure. Using the defining relations for the creation and annihilation operators, Eqs. (2.104) and (2.105), and the definition of the inner product (2.107), we obtain a differential equation for $\mu(z, \bar{z})$:

$$\frac{i\gamma}{m_{\rm Pl}^2}(pr)k\frac{\partial \mu}{\partial \bar{z}_{rp}} = z_{rp} . \tag{2.108}$$

This can be integrated to give

$$\mu(z, \bar{z}) = \int d\mathbf{k} \frac{m_{\rm Pl}^2}{k} \sum_{rp} \frac{pr}{i\gamma} z_{rp}(\mathbf{k}) \bar{z}_{rp}(\mathbf{k}) , \tag{2.109}$$

which fixes $\langle \Phi_1 | \Phi_2 \rangle$. The vacuum of this representation is defined by $g_{rp}\Phi_0 = 0$ which gives

$$\Phi_0 = \langle z | 0 \rangle = 1 , \tag{2.110}$$

and particle states are monomials in the respective variables,

$$\Phi_n = \langle z | n \rangle \propto (g^{\dagger}_{rp})^n \Psi_0 = z^n_{rp} . \tag{2.111}$$

These states are not normalisable for $i\gamma = pr$, as in this case the measure is positive and the exponential in (2.107) blows up. Hence, these states should be removed from the physical Hilbert space and therefore their associated operators g_{rp} should not appear in the Hamiltonian. For $\gamma = i$, this only leaves two physical modes $g_R^{ph} = g_{R+}$ and $g_L^{ph} = g_{L-}$.

For the SD connection we therefore obtain the physical Hamiltonian

$$\mathcal{H}^{ph}_{eff} \approx m_{\rm Pl}^2 \int d\mathbf{k} \, (g_L^{ph} g_L^{ph\dagger} + g_R^{ph\dagger} g_R^{ph}) . \tag{2.112}$$

This looks like the standard Hamiltonian for a harmonic oscillator, with the difference that only the left handed graviton needs to be normal ordered and produces a vacuum energy. For the ASD connection only the right handed graviton produces vacuum energy. Left and right handed gravitons are not on the same footing, and the theory is chiral. We will explore this chirality in more detail after finding the graviton operators for general γ.

2.3.4.2 Hamiltonian for Complex Values of γ

Let us focus on the general Hamiltonian (inside the horizon) in terms of modes again, Eq. (2.97). We need to identify linear combinations of metric and connection that can act as graviton operators, equivalent to g_{rp} and $g^{\dagger}_{rp}$ for $\gamma = \pm i$. We want to end up with two physical operators corresponding to the two independent polarizations, however initially there should be four different operators. Two of them will be zero on-shell, representing the unphysical modes, while the other two should commute with them [c.f. Eq. (2.103)] and reduce to metric variables on-shell.

To find the general expression, consider the graviton operators for $\gamma = \pm i$ and find linear combinations of them that satisfy these conditions. After some algebraic manipulations that make use of the on-shell conditions (2.85) and (2.86), we can identify suitable operators:

$$G_{r\mathcal{P}_+} = \frac{(r - i\gamma)g_{r+} - (r + i\gamma)g_{r-}}{-2\gamma i}, \tag{2.113}$$

$$G_{r\mathcal{P}_-} = \frac{(r + i\gamma)g_{r+} - (r - i\gamma)g_{r-}}{-2\gamma i}, \tag{2.114}$$

where the new index $\mathcal{P} = \mathcal{P}_+, \mathcal{P}_-$ labels physical and non-physical modes. This notation is used to avoid confusions with $p = \pm 1$ used for positive and negative frequencies, and except for the cases of $\gamma = \pm i$, the two indices do not align. Using the on-shell conditions, we find that $G_{r\mathcal{P}_-} \approx 0$ and $G_{r\mathcal{P}_+} \approx 2rke_r$ as required, and you can check that their commutator is zero. The index $\mathcal{P} = \mathcal{P}_+ = 1$ therefore denotes physical modes, which reduce to the metric classically (and quantum mechanically will have positive energy and norm), and $\mathcal{P} = \mathcal{P}_- = -1$ denotes modes that vanish on-shell (and quantum mechanically will have negative energy and norm).

We can use the expressions in Eqs. (2.99) to (2.102) to write the new operators $G_{r\mathcal{P}}$ in terms of metric and connection variables. We can find expressions for the creation operators by demanding that they are hermitian conjugates of the annihilation operators once the reality conditions (2.81) and (2.82) are imposed. The operators and their commutators are listed in Table 2.2.

The Hamiltonian (2.97) can be written in terms of the new graviton operators as

$$\begin{aligned}\mathcal{H}_{eff} = \frac{m_{\text{Pl}}^2}{2}\int d^3k \sum_r &-(1 + i\gamma r)G_{r\mathcal{P}_+}(\mathbf{k})G_{r\mathcal{P}_-}(-\mathbf{k}) - (1 - i\gamma r)G_{r\mathcal{P}_-}(\mathbf{k})G_{r\mathcal{P}_+}(-\mathbf{k})\\ &+(1 + i\gamma r)G_{r\mathcal{P}_+}(\mathbf{k})G^{\dagger}_{r\mathcal{P}_+}(\mathbf{k}) + (1 - i\gamma r)G^{\dagger}_{r\mathcal{P}_+}(\mathbf{k})G_{r\mathcal{P}_+}(\mathbf{k})\end{aligned}$$

$$+(1-i\gamma r)G_{r\mathcal{P}_-}(\mathbf{k})G^\dagger_{r\mathcal{P}_-}(\mathbf{k})+(1+i\gamma r)G^\dagger_{r\mathcal{P}_-}(\mathbf{k})G_{r\mathcal{P}_-}(\mathbf{k})$$
$$-(1-i\gamma r)G^\dagger_{r\mathcal{P}_+}(\mathbf{k})G^\dagger_{r\mathcal{P}_-}(-\mathbf{k})-(1+i\gamma r)G^\dagger_{r\mathcal{P}_-}(\mathbf{k})G^\dagger_{r\mathcal{P}_+}(-\mathbf{k})\,. \tag{2.115}$$

This is the generalization of Eq. (2.98). As before, there are too many graviton states as well as unphysical pair production terms. They all vanish on shell where the operator corresponding to $\mathcal{P}_-$ is zero. We can now set up the Hilbert space, fixing the inner product by requiring that the operators in Table 2.2 are indeed hermitian conjugates of one another.

Again, we use a holomorphic representation which diagonalises $G^\dagger_{r\mathcal{P}}$, i.e.:

$$\langle z|G^\dagger_{r\mathcal{P}}|\Phi\rangle = z_{r\mathcal{P}}\langle z|\Phi\rangle\,. \tag{2.116}$$

The commutation relations in Table 2.2 determine the action of the annihilation operators,

$$\langle z|G_{r\mathcal{P}}|\Phi\rangle = \mathcal{P}\frac{k}{m^2_{\rm Pl}}\frac{\partial}{\partial z_{r\mathcal{P}}}\langle z|\Phi\rangle\,. \tag{2.117}$$

This is formally very similar to the case $\gamma=\pm i$, but note that the variables $z_{r\mathcal{P}}$ are not the same as before. Using the definition of the inner product Eq. (2.107), and the same formal condition $\langle\Phi_1|G^\dagger_{r\mathcal{P}}|\Phi_2\rangle = \overline{\langle\Phi_2|G_{r\mathcal{P}}|\Phi_1\rangle}$, we arrive at an expression for the measure:

$$\mu(z,\bar z) = \int d\mathbf{k}\frac{m^2_{\rm Pl}}{k}\sum_{r\mathcal{P}}\mathcal{P}z_{r\mathcal{P}}(\mathbf{k})\bar z_{r\mathcal{P}}(\mathbf{k})\,. \tag{2.118}$$

The vacuum state

$$\Phi_0=\langle z|0\rangle = 1\,, \tag{2.119}$$

and the particle states

$$\Phi_n=\langle z|n\rangle \propto (G^\dagger_{r\mathcal{P}})^n\Psi_0 = z^n_{r\mathcal{P}}\,, \tag{2.120}$$

have the same form as before (but are defined in terms of new variables $z_{r\mathcal{P}}$). The measure implies that states with $\mathcal{P}=\mathcal{P}_-=-1$ are not normalisable and the

Table 2.2 Physical and unphysical graviton modes

Physical $\mathcal{P}=\mathcal{P}_+=1$	Unphysical $\mathcal{P}=\mathcal{P}_-=-1$
$G_{r\mathcal{P}_+}=\frac{-r}{i\gamma}(\tilde a_{r+}-k(r+i\gamma)\tilde e_{r+})$	$G_{r\mathcal{P}_-}=\frac{-r}{i\gamma}(\tilde a_{r+}-k(r-i\gamma)\tilde e_{r+})$
$G^\dagger_{r\mathcal{P}_+}=\frac{r}{i\gamma}(\tilde a^\dagger_{r-}-k(r-i\gamma)\tilde e^\dagger_{r-})$	$G^\dagger_{r\mathcal{P}_-}=\frac{r}{i\gamma}(\tilde a^\dagger_{r-}-k(r+i\gamma)\tilde e^\dagger_{r-})$
$\left[G_{r\mathcal{P}_+}(\mathbf{k}),G^\dagger_{s\mathcal{P}_+}(\mathbf{k}')\right]=\frac{k}{m^2_{\rm Pl}}\delta_{rs}\delta(\mathbf{k}-\mathbf{k}')$	$\left[G_{r\mathcal{P}_-}(\mathbf{k}),G^\dagger_{s\mathcal{P}_-}(\mathbf{k}')\right]=-\frac{k}{m^2_{\rm Pl}}\delta_{rs}\delta(\mathbf{k}-\mathbf{k}')$

operators corresponding to $\mathcal{P}_-$ should be removed from the Hamiltonian. The physical Hamiltonian for a general value of γ is therefore:

$$\mathcal{H}_{eff}^{ph} \approx \frac{m_{\rm Pl}^2}{2} \int d\mathbf{k} \sum_r [G_r^{ph} G_r^{ph\dagger}(1 + ir\gamma) + G_r^{ph\dagger} G_r^{ph}(1 - ir\gamma)], \qquad (2.121)$$

where $G_r^{ph} = G_r \mathcal{P}_+$.

2.3.4.3 Vacuum Energy

Only the first term in the physical Hamiltonian (2.121) needs to be normal ordered, using the commutation relation in Table 2.2. This leads to a chiral (r-dependent) term corresponding to the vacuum energy, $V_r \propto 1 + ir\gamma$. The asymmetry in the vacuum energy between the right- and left-handed gravitons is given by

$$\frac{V_R - V_L}{V_R + V_L} = i\gamma . \qquad (2.122)$$

This equation is valid for any complex γ. There are a few points of interest to note. If γ is purely imaginary and $|\gamma| > 1$, the vacuum energy $V_r \propto 1 + ir\gamma$ of one of the modes becomes negative. Negative vacuum energy is often associated with fermionic degrees of freedom [60], but this will not be investigated further here.

More importantly, if γ has a real part the VE for each r is complex. When right and left helicities are added together, however, we simply obtain $V_R + V_L \propto 1 + i\gamma + 1 - i\gamma = 2$, so the total vacuum energy is indeed real.

The reason we obtain a chiral, complex vacuum energy is because the Hamiltonian is not hermitian before normal ordering: Although it is real on-shell for any value of γ (which does not appear in any on-shell expressions) and the graviton operators themselves are hermitian, unless γ is imaginary, taking the hermitian conjugate of the perturbative physical Hamiltonian (2.121) does not yield $\mathcal{H}^\dagger = \mathcal{H}$.

Hermiticity is restored after normal ordering, when γ drops out of the Hamiltonian and is only present in the vacuum energy term. As the latter is not physically measurable (and when coupled to the Einstein equations, we need to consider the total which is indeed real), this result might not be too concerning. However it might also imply that it is more physical to consider only a purely imaginary γ or that we should use a symmetric ordering for the Hamiltonian: When we first defined the Hamiltonian in Eq. (2.42), we picked an ordering of the form EEF (the field strength contains connection terms, which do not commute with metric terms). Knowing which ordering in quantum mechanics is "correct" is an issue which can ultimately only be resolved by experiment. It can be checked that using an EFE or $\frac{1}{2}(EEF + FEE)$ ordering would satisfy $\mathcal{H} = \mathcal{H}^\dagger$ on and off-shell, for any value of γ. In this case there would be no chirality in the vacuum energy. However, note that we would obtain the same graviton operators regardless of ordering, and as will see now, chirality will still be present in the vacuum fluctuations.

2.3.5 Chiral Vacuum Fluctuations

The central gravitational variable in the Ashtekar formalism is the connection, not the metric, which can be seen from the Holst action (2.22). Therefore, the power spectrum of tensor perturbations should be derived from the (TT-projected) perturbations of the connection as opposed to the metric. As in the second order formalism, the Ashtekar tensor perturbations will have an effect on the CMB fluctuations, especially on the polarization. We will not need to worry about the exact normalization of the tensor fluctuations, as we are mainly interested to see whether the complex nature of the connection will play a role.

The analogous expression to the tensor power spectrum (1.65) is given by

$$\langle 0|A_r^\dagger(\mathbf{k})A_r(\mathbf{k}')|0\rangle = P_r(k)\delta(\mathbf{k}-\mathbf{k}') \ , \tag{2.123}$$

where $A_r(\mathbf{k})$ represents Fourier space connection variables with handedness r, i.e.

$$A_r(\mathbf{k}) = \tilde{a}_{r+}(\mathbf{k})e^{-ik\cdot x} + \tilde{a}_{r-}^\dagger(\mathbf{k})e^{ik\cdot x} \ . \tag{2.124}$$

Note that we could have picked a different ordering in the 2-point function (2.123), so in general we have to consider

$$A^\dagger A \to \alpha A^\dagger A + \beta A A^\dagger \ , \tag{2.125}$$

with $\alpha + \beta = 1$ and $\alpha, \beta > 0$. As opposed to the vacuum energy, we will see that the power spectrum (2.123), being a measurable variance, is always real and positive.

To compute the physical power spectrum, we need to relate the connection variables to the physical graviton modes labelled by $\mathcal{P}_+$ in Table 2.2. As we need to go on-shell to define physical states, we can use conditions (2.85) and (2.86) to express the metric variables in terms of the connection:

$$\tilde{e}_{r+} = \frac{\tilde{a}_{r+}}{k(r-i\gamma)} \ , \quad \tilde{e}_{r-}^\dagger = \frac{\tilde{a}_{r-}^\dagger}{k(r+i\gamma)} \ . \tag{2.126}$$

These relations can be subsituted into the equations for $G_{r\mathcal{P}_+}^\dagger$, $G_{r\mathcal{P}_+}$ in Table 2.2, which gives expressions for the physical connection modes a_{r+}^{ph} and $a_{r-}^{ph\dagger}$. The remaining modes can be obtained by taking hermitian conjugates (as we are on-shell, the reality conditions have been imposed). We find

$$a_{r+}^{ph} = \frac{r-i\gamma}{2r}G_{r\mathcal{P}_+} \ , \tag{2.127}$$

$$a_{r+}^{ph\dagger} = \frac{r+i\gamma^*}{2r}G_{r\mathcal{P}_+}^\dagger \ , \tag{2.128}$$

$$a_{r-}^{ph} = \frac{r-i\gamma^*}{2r}G_{r\mathcal{P}_+} \ , \tag{2.129}$$

$$a_{r-}^{ph\dagger} = \frac{r+i\gamma}{2r} G_{r\mathcal{P}_+}^{\dagger} \,. \tag{2.130}$$

We can see that the physical connection modes depend solely on the graviton operators, so they will be the same for any ordering of the Hamiltonian. Plugging these expressions into (2.124) we obtain for the two connection helicity states:

$$A_r^{ph}(\mathbf{k}) = \frac{r-i\gamma}{2r} G_{r\mathcal{P}_+}(\mathbf{k}) e^{-ik\cdot x} + \frac{r+i\gamma}{2r} G_{r\mathcal{P}_+}^{\dagger}(\mathbf{k}) e^{ik\cdot x} \,,$$
$$A_r^{ph\dagger}(\mathbf{k}) = \frac{r-i\gamma^*}{2r} G_{r\mathcal{P}_+}(\mathbf{k}) e^{-ik\cdot x} + \frac{r+i\gamma^*}{2r} G_{r\mathcal{P}_+}^{\dagger}(\mathbf{k}) e^{ik\cdot x} \,.$$

This means that the power spectrum (2.123) is given by (using $G_{r\mathcal{P}_+}(\mathbf{k}')|0\rangle = 0$)

$$\langle 0|A_r^{ph\dagger}(\mathbf{k}) A_r^{ph}(\mathbf{k}')|0\rangle = P_r(\gamma) \langle 0|G_{r\mathcal{P}_+}(\mathbf{k}) G_{r\mathcal{P}_+}^{\dagger}(\mathbf{k}')|0\rangle \,. \tag{2.131}$$

We could eliminate the expectation value of graviton operators by using their commutator to give us an expression in terms of delta functions. However, we are only interested in the chiral dependence of the power spectrum P_r, which is given by

$$P_r(\gamma) = \frac{(r+i\gamma)(r-i\gamma^*)}{4} = \frac{1-2\gamma_I r + |\gamma|^2}{4} \,. \tag{2.132}$$

If $\gamma_I r < 0$, $P_r(\gamma)$ is obviously positive. Otherwise,

$$P_r(\gamma) \propto 1 - 2|\gamma_I| + \gamma_I^2 + \gamma_R^2 = (1-|\gamma_I|)^2 + \gamma_R^2 \,, \tag{2.133}$$

which is also positive for any complex γ. Therefore, the 2-point function is indeed always real and positive, as required. The chiral asymmetry in the power spectrum can be written as

$$\frac{P_R - P_L}{P_R + P_L} = -\frac{2\gamma_I}{1+|\gamma|^2} \,, \tag{2.134}$$

or, for a general ordering of the 2-point function as in (2.125),

$$\frac{P_R - P_L}{P_R + P_L} = \frac{2(\beta-\alpha)\gamma_I}{1+|\gamma|^2} \,. \tag{2.135}$$

The chirality in the power spectrum of tensor fluctuations is the main new result of this work, and a big difference to the standard second order formalism described in Sect. 1.2.5 (which corresponds to the limit $|\gamma| \to \infty$, for which the Holst action reduces to the Palatini action).

We can see that if γ was purely real there would be no asymmetry in the vacuum fluctuations for right and left gravitons. The chirality is related to the fact that for a γ with an imaginary part the connection is a complex field and therefore we must expand it in terms of graviton and anti-graviton modes. Note, however, that a real

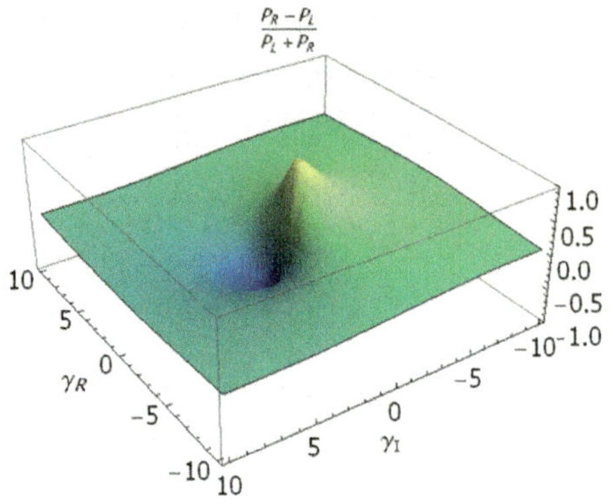

Fig. 2.1 Power spectrum asymmetry as a function of a generally complex Immirzi parameter γ

part in the Immirzi parameter does affect the absolute value of the asymmetry due to the factor $|\gamma|$ in the denominator of (2.134). We can also see that for a completely symmetric ordering of the 2-point function, $\alpha = \beta$, the RHS of Eq. (2.135) is zero. Hence, even if the Ashtekar formalism was the correct description of gravity, we would not obtain a chiral power spectrum if γ was real or the ordering symmetric. Not measuring chirality would therefore not be able to rule out the theory.

We can plot the power spectrum asymmetry (2.134) against the real and imaginary parts of γ, see Fig. 2.1. It is obviously antisymmetric in γ_I, and the minimum and maximum are at $\gamma = \pm i$ respectively which are the values that correspond to a SD/ASD connection. They display the maximum chirality because the Palatini action can naturally be split into a SD and ASD part [19]. The axis $\gamma_I = 0$ corresponds to a real γ and therefore displays no asymmetry.

2.3.5.1 Measuring a Chiral Tensor Spectrum

As was mentioned in Sect. 1.3.2, in the absence of parity violation, the TB power spectrum of the CMB would be zero. In the situation we have just considered, the chirality of the power spectrum (2.134) breaks parity. The effect of parity violation on the CMB power spectra was investigated in [61]. It was found that the ratio between the quadrupole of the TB correlator (zero in standard cosmological approaches) and the BB correlator is given by

$$\frac{C_2^{\mathrm{TB}}}{C_2^{\mathrm{BB}}} \approx f_{\mathrm{PB}}\alpha_2\,, \tag{2.136}$$

where $\alpha_2 \approx 200$ parametrises the relative strength between the TB and BB spectra and $f_{\mathrm{PB}} = 2\frac{P_R - P_L}{P_R + P_L}$ is the parity breaking parameter, which is zero if no chirality is present. In our case, we therefore find

$$\frac{C_2^{\mathrm{TB}}}{C_2^{\mathrm{BB}}} \approx 800\frac{(\beta - \alpha)\gamma_I}{1 + |\gamma|^2}\,, \tag{2.137}$$

for the ratio of tensor induced TB and BB quadrupole modes. Not only would chirality render the TB correlator non-zero, it would also be easier to detect TB rather than BB correlation ($C_2^{\rm TB} > C_2^{\rm BB}$) for a wide range of values of γ, given approximately by

$$\frac{1}{800} < |\gamma| < 800\,. \tag{2.138}$$

BICEP2 has recently detected B-modes [62] that might have arisen due to tensor perturbations from inflation, however we do not yet have tensor power spectra over a large number of multipoles as the experiment only took data from a small patch of sky. Although there was no hint of parity violation in their analysis so far, this might change once the full power spectrum becomes available. It will therefore be possible in the near future to constrain the model I have described. If the TB correlator is consistent with zero, we know that for Ashtekar gravity to be correct, γ must be either quite far from the range in Eq. (2.138) or real. If a chirality was detected, on the other hand, it could indeed have originated from this mechanism.

2.3.6 A Purely Real γ

Before I conclude, let us quickly consider the case of a purely real theory for which $\Im(\gamma) = 0$. Although it will turn out that we can take the limit $\Im(\gamma) \to 0$ in all of our main results to obtain the answer in the real theory, it is not initially obvious why this would work, as a real theory is very different from a complex one. I will describe the main differences and show why our results are still well defined in the real case.

A purely real theory would require Fourier mode expansions using operators a_r and e_r without a p index, as there is no need to consider separate sets of creation and annihilation operators. We therefore would only get two modes for each $\mathbf{k}$ and r as usual in the second order theory. As we ignore the p index, what used to be reality conditions in the complex theory, where we related modes with different p, are now just operator conditions, $\tilde{e}_{r+} = \tilde{e}_{r-}$ and $\tilde{a}_{r+} = \tilde{a}_{r-}$. Similarly, the commutation relations (2.95) have one less index and must be replaced by

$$[\tilde{a}_r(\mathbf{k}), \tilde{e}_s^\dagger(\mathbf{k}')] = -i\frac{\gamma}{2m_{\rm Pl}^2}\delta_{rs}\delta(\mathbf{k}-\mathbf{k}')\,. \tag{2.139}$$

The Hamiltonian, on the other hand, will still have the same form, as $p = -1$ modes always appear with a dagger and $p = +1$ modes without, see Eq. (2.97). This enables us to define the same physical and unphysical graviton operators as before, however without a p index on the RHS, e.g.

$$G_{r\mathcal{P}_+} = \frac{-r}{i\gamma}(\tilde{a}_r - k(r+i\gamma)\tilde{e}_r)\,. \tag{2.140}$$

Note that as opposed to the complex case, were the graviton operators were only hermitian conjugates of each other after the reality conditions had been imposed, for the real theory the reality conditions are satisfied by the metric and connection operators. Therefore, $G_{r\mathcal{P}}$ and $G^{\dagger}_{r\mathcal{P}}$ are automatically conjugates of one another, which can be trivially seen from their definitions.

We still have a non-physical mode, however, which can be eliminated by imposing the torsion free condition which relates a_r to e_r. As before, we can define a holomorphic representations and an inner product, which will show that the non-physical modes have negative energy and should therefore be excluded. Our Hamiltonian and Hilbert space will therefore have the same structure as for a general complex γ. Hence, the real theory can be viewed as the limit $\Im(\gamma) \to 0$ in the sections above.

2.4 Conclusions

I have shown that using the Ashtekar formalism in cosmological perturbation theory leads to a number of interesting results.

Classically, rederiving the second order Hamiltonian corresponding to tensor perturbations is far from trivial. We saw that we need to take boundary terms into account, as well as regard the change from the full Ashtekar variables to the perturbations as a canonical transformation in order to arrive at the correct form of the Hamiltonian. I was then able to reproduce the standard result for the equation of motion of tensor modes, as obtained in the second order formalism.

On the quantum mechanical front there were several novelties. First of all, the fact that the connection is complex makes the exercise a lot more involved than in the usual case. We need to expand the fields in terms of positive and negative frequency operators, which are related by reality conditions. These are not supposed to be imposed on the operators, but only at the very end when choosing the inner product of the Hilbert space. We can write the Hamiltonian in terms of graviton creation and annhilation operators, which are linear combinations of metric and connection. When fixing the inner product, we find that half of the operators are unphysical, demonstrated by them being zero-on shell, when the torsion free condition relating metric and connection is imposed. This also gets rid of unphysical coupling terms between $\mathbf{k}$ and $-\mathbf{k}$ in the Hamiltonian.

As the connection is complex, so is the dynamical, perturbed Hamiltonian. This is not a problem as we ensure actual observables are real by requiring the Hamiltonian to be hermitian through the choice of inner product, at least after normal ordering. The complexity of the Hamiltonian is, however, the origin of the chiral effects we observe.

Before normal ordering, if $\gamma_R \neq 0$, the Hamiltonian is not hermitian, which results in an imaginary vacuum energy for each helicity. Non-hermitian Hamiltonians have been studied before [63] and are not necessarily regarded as problematic. In our case, the total vacuum energy for both helicities is real, and therefore the non-hermitian nature might not be physically significant.

The main result of this chapter is the chiral power spectrum of tensor perturbations, which is described in terms of perturbed connection variables. This chirality is present as long as γ is not purely real, and the strongest effect occurs for the SD/ASD connection for which $\gamma = \pm i$. The chirality in the power spectrum is a novelty compared to the standard second order formalism, and demonstrates that using different variables to describe spacetime does not necessarily lead to equivalent results.

A chiral graviton would break parity and therefore lead to a non-zero TB correlator, which can be probed by CMB measurements. As the Planck collaboration will release their polarization results later this year, it is only a matter of time until the full power spectrum can be obtained, which will enable us to constrain the value of the Immirzi parameter.

Although gravitational chirality can be produced in other ways [60, 64, 65], the mechanism presented here is by far the simplest. If a chiral tensor power spectrum was to be observed, it would hint at the Ashtekar formalism being the correct fundamental description of gravity.

References

1. J. Polchinski, *String Theory* (*Cambridge Monographs on Mathematical Physics*) (Cambridge University Press, 1998)
2. L. Bethke, J. Magueijo, Inflationary tensor fluctuations, as viewed by Ashtekar variables and their imaginary friends. Phys. Rev. **D84**, 024014 (2011). http://xxx.lanl.gov/abs/1104.1800
3. L. Bethke, J. Magueijo, Chirality of tensor perturbations for complex values of the Immirzi parameter. Class. Quant. Grav. **29**, 052001 (2012). http://xxx.lanl.gov/abs/1108.0816
4. A. Zee, *Quantum Field Theory in a Nutshell* (Princeton University Press, 2010)
5. P. Dirac, *Principles of Quantum Mechanics* (Oxford University Press, 1982)
6. M.E. Peskin, D.V. Schroeder, *An Introduction to Quantum Field Theory* (Westview Press Inc., 1995)
7. R. Feynman, Space-time approach to nonrelativistic quantum mechanics. Rev. Mod. Phys. **20**, 367–387 (1948)
8. M.B. Green, J.H. Schwarz, E. Witten, *Superstring Theory* (*Cambridge Monographs on Mathematical Physics*) (Cambridge University Press, 1987)
9. C. Rovelli, *Quantum Gravity* (*Cambridge Monographs on Mathematical Physics*) (Cambridge University Press, 2004)
10. R.L. Arnowitt, S. Deser, C.W. Misner, Canonical variables for general relativity. Phys. Rev. **117**, 1595–1602 (1960)
11. H. Sahlmann, Loop quantum gravity—a short review. http://xxx.lanl.gov/abs/1001.4188
12. B.S. DeWitt, Quantum theory of gravity. 1. the canonical theory. Phys. Rev. **160**, 1113–1148 (1967)
13. A. Ashtekar, New variables for classical and quantum gravity. Phys. Rev. Lett. **57**, 2244–2247 (1986)
14. A. Ashtekar, New Hamiltonian formulation of general relativity. Phys. Rev. **D36**, 1587–1602 (1987)
15. C. Rovelli, L. Smolin, Knot theory and quantum gravity. Phys. Rev. Lett. **61**, 1155 (1988)
16. C. Rovelli, L. Smolin, Loop space representation of quantum general relativity. Nucl. Phys. **B331**, 80 (1990)
17. K.G. Wilson, Confinement of quarks. Phys. Rev. **D10**, 2445–2459 (1974)
18. T. Jacobson, L. Smolin, Nonperturbative quantum geometries. Nucl. Phys. **B299**, 295 (1988)

19. T. Thiemann, Modern canonical quantum general relativity. http://xxx.lanl.gov/abs/gr-qc/0110034
20. C. Rovelli, L. Smolin, Spin networks and quantum gravity. Phys. Rev. **D52**, 5743–5759 (1995). http://xxx.lanl.gov/abs/gr-qc/9505006
21. J.C. Baez, Spin network states in gauge theory. Adv. Math. **117**, 253–272 (1996). http://xxx.lanl.gov/abs/gr-qc/9411007
22. A. Ashtekar, J. Lewandowski, Quantum theory of geometry. 1: area operators. Class. Quant. Grav. **14**, A55–A82 (1997). http://xxx.lanl.gov/abs/gr-qc/9602046
23. A. Ashtekar, J. Lewandowski, Quantum theory of geometry. 2. volume operators. Adv. Theor. Math. Phys. **1**, 388–429 (1998). http://xxx.lanl.gov/abs/gr-qc/9711031
24. A. Ashtekar, J. Lewandowski, Projective techniques and functional integration for gauge theories. J. Math. Phys. **36**, 2170–2191 (1995). http://xxx.lanl.gov/abs/gr-qc/9411046
25. A. Ashtekar, J. Lewandowski, Differential geometry on the space of connections via graphs and projective limits. J. Geom. Phys. **17**, 191–230 (1995). http://xxx.lanl.gov/abs/hep-th/9412073
26. H.A. Morales-Tecotl, C. Rovelli, Fermions in quantum gravity. Phys. Rev. Lett. **72**, 3642–3645 (1994). http://xxx.lanl.gov/abs/gr-qc/9401011
27. J.C. Baez, K.V. Krasnov, Quantization of diffeomorphism invariant theories with fermions. J. Math. Phys. **39**, 1251–1271 (1998). http://xxx.lanl.gov/abs/hep-th/9703112
28. L. Modesto, C. Rovelli, Particle scattering in loop quantum gravity. Phys. Rev. Lett. **95**, 191301 (2005). http://xxx.lanl.gov/abs/gr-qc/0502036
29. C. Rovelli, Graviton propagator from background-independent quantum gravity. Phys. Rev. Lett. **97**, 151301 (2006). http://arxiv.org/abs/gr-qc/0508124
30. C. Rovelli, Loop quantum gravity. Living Rev. Rel. **1**, 1 (1998). http://xxx.lanl.gov/abs/gr-qc/9710008
31. C. Rovelli, Black hole entropy from loop quantum gravity. Phys. Rev. Lett. **77**, 3288–3291 (1996). http://xxx.lanl.gov/abs/gr-qc/9603063
32. M. Bojowald, Loop quantum cosmology. Living Rev. Rel. **11**, 4 (2008)
33. A. Ashtekar, M. Bojowald, J. Lewandowski, Mathematical structure of loop quantum cosmology. Adv. Theor. Math. Phys. **7**, 233–268 (2003). http://xxx.lanl.gov/abs/gr-qc/0304074
34. M. Bojowald, Inflation from quantum geometry. Phys. Rev. Lett. **89**, 261301 (2002). http://xxx.lanl.gov/abs/gr-qc/0206054
35. M. Bojowald, Absence of singularity in loop quantum cosmology. Phys. Rev. Lett. **86**, 5227–5230 (2001). http://xxx.lanl.gov/abs/gr-qc/0102069
36. M. Bojowald, Quantum nature of cosmological bounces. Gen. Rel. Grav. **40**, 2659–2683 (2008). http://xxx.lanl.gov/abs/0801.4001
37. M.P. Reisenberger, C. Rovelli, 'Sum over surfaces' form of loop quantum gravity. Phys. Rev. **D56**, 3490–3508 (1997). http://xxx.lanl.gov/abs/gr-qc/9612035
38. J.C. Baez, Spin foam models. Class. Quant. Grav. **15**, 1827–1858 (1998). http://xxx.lanl.gov/abs/gr-qc/9709052
39. J. Ambjorn, J. Jurkiewicz, R. Loll, Reconstructing the universe. Phys. Rev. **D72**, 064014 (2005). http://xxx.lanl.gov/abs/hep-th/0505154
40. L. Bombelli, J. Lee, D. Meyer, R. Sorkin, Space-time as a causal set. Phys. Rev. Lett. **59**, 521–524 (1987)
41. R.D. Sorkin, Causal sets: discrete gravity. http://xxx.lanl.gov/abs/gr-qc/0309009
42. S. Carroll, *Spacetime and Geometry: An Introduction to General Relativity* (Benjamin Cummings, 2003)
43. P. Dona, S. Speziale, Introductory lectures to loop quantum gravity. http://xxx.lanl.gov/abs/1007.0402
44. R.M. Wald, *General Relativity* (The University of Chicago Press, 1984)
45. S. Holst, Barbero's Hamiltonian derived from a generalized Hilbert-Palatini action. Phys. Rev. **D53**, 5966–5969 (1996). http://xxx.lanl.gov/abs/gr-qc/9511026
46. G. Gibbons, S. Hawking, Action integrals and partition functions in quantum gravity. Phys. Rev. **D15**, 2752–2756 (1977)

47. T. Regge, C. Teitelboim, Role of surface integrals in the hamiltonian formulation of general relativity. Annals Phys. **88**, 286 (1974)
48. A. Ashtekar, J. Engle, D. Sloan, Asymptotics and Hamiltonians in a first order formalism. Class. Quant. Grav. **25**, 095020 (2008). http://xxx.lanl.gov/abs/0802.2527
49. A. Ashtekar, C. Rovelli, L. Smolin, Gravitons and loops. Phys. Rev. **D44**, 1740–1755 (1991). http://xxx.lanl.gov/abs/hep-th/9202054
50. R. Courant, D. Hilbert, *Methods of Mathematical Physics, Volume I* (Wiley-Interscience, 1962)
51. H. Goldstein, *Classical Mechanics* (Addison-Wesley, Boston, 1980)
52. L. Freidel, L. Smolin, The linearization of the Kodama state. Class. Quant. Grav. **21**, 3831–3844 (2004). http://xxx.lanl.gov/abs/hep-th/0310224
53. L. Grishchuk, Y. Sidorov, Squeezed quantum states of relic gravitons and primordial density fluctuations. Phys. Rev. **D42**, 3413–3421 (1990)
54. T.K. Misner, C. and J. Wheeler, Gravitation
55. A. Ashtekar, A note on helicity and selfduality. J. Math. Phys. **27**, 824–827 (1986)
56. S. Weinberg, Photons and gravitons in perturbation theory: derivation of Maxwell's and Einstein's equations. Phys. Rev. **138**, B988–B1002 (1965)
57. V. Bargmann, On a Hilbert space of analytic functions and an associated integral transform. 1. Commun. Pure Appl. Math. **14**, 187–214 (1961)
58. R. Gambini, J. Pullin, *Loops, Knots, Gauge Theories and QuantumGravity* (Cambridge University Press, Gauge Theories and Quantum Gravity, 1996)
59. A. Ashtekar, R.S. Tate, An algebraic extension of Dirac quantization: examples. J. Math. Phys. **35**, 6434–6470 (1994). http://xxx.lanl.gov/abs/gr-qc/9405073
60. S.H. Alexander, G. Calcagni, Quantum gravity as a fermi liquid. Found. Phys. **38**, 1148–1184 (2008). http://xxx.lanl.gov/abs/0807.0225
61. C.R. Contaldi, J. Magueijo, L. Smolin, Anomalous CMB polarization and gravitational chirality. Phys. Rev. Lett. **101**, 141101 (2008). http://xxx.lanl.gov/abs/0806.3082
62. BICEP2 Collaboration, P.A.R. Ade et al., BICEP2 I: detection of b-mode polarization at degree angular scales. http://xxx.lanl.gov/abs/1403.3985
63. C.M. Bender, Making sense of non-Hermitian Hamiltonians. Rept. Prog. Phys. **70**, 947 (2007). http://xxx.lanl.gov/abs/hep-th/0703096
64. S.H. Alexander, Isogravity: toward an electroweak and gravitational unification. http://xxx.lanl.gov/abs/0706.4481
65. S. Mercuri, Modifications in the spectrum of primordial gravitational waves induced by instantonic fluctuations. Phys. Rev. **D84**, 044035 (2011). http://xxx.lanl.gov/abs/1007.3732

Chapter 3
Anisotropic Gravitational Wave Background from Massless Preheating

Reheating is one of the least understood periods in the early universe. While nearly all the elementary particles we observe must have been produced during this period, we still do not know for sure how the process occurred. Although reheating might have an effect on the curvature perturbation [1], the main constraint to date comes from the abundance of light elements which give bounds on the reheating temperature after thermalisation [2]. However, as was shown in Sect. 1.4, the detailed preheating dynamics strongly depend on the underlying model of inflation. Therefore, studying observables that were affected by the reheating process would give us insight into this period as well as inflation.

Gravitational waves are an ideal candidate to probe the period of reheating, and therefore inflation, further. As mentioned in Sect. 1.5.2, they will be produced in large quantities during preheating due to the presence of time-varying matter inhomogeneities, and their spectrum will peak at a scale that is characteristic of the preheating dynamics.

If the background of GWs from preheating was to be measured, it would therefore provide information about the inflaton potential, as well as the couplings of the inflaton to other matter fields, which cannot be easily be obtained in other ways. Unlike the CMB fluctuations, GWs from preheating decouple right upon production (below the Planck scale), due to their weak interaction with other matter [3]. Therefore, they do not evolve on their journey towards us, and retain their spectral shape and frequency (except for a redshift due to expansion), giving us a direct snapshot of the very early universe. Even though it is currently not possible to directly measure GW backgrounds from preheating, which peak at very high frequencies [4], it is still important to characterise them, as they might become a vital tool of observational Cosmology in the future.

In this section I will discuss GW production from massless preheating, in the presence of a light scalar field. As this is a non-linear process, the problem will naturally have to be solved numerically. I will show that this model leads to an anisotropic background of GWs today, with relative fluctuations of the order of 1 %, as was presented in [5, 6]. Such anisotropies could arise in any preheating scenario

© Springer International Publishing Switzerland 2015

L.B. Bethke, *Exploring the Early Universe with Gravitational Waves*,
Springer Theses, DOI 10.1007/978-3-319-17449-5_3

where a light scalar field is present, providing a novel way to distinguish between different inflationary models.

I will start by describing gravitational wave production from preheating and how it can be studied numerically in Sect. 3.1, with particular focus on massless preheating. I will also explain how to relate the frequency and energy density of the produced GWs to their values today.

In Sect. 3.2, I will argue why the presence of a light scalar field, which acquires a scale-invariant spectrum of perturbations during inflation, would result in anisotropies in the GW background. The numerical algorithm and the parameters used in the simulations are the topic of Sect. 3.3, where I will show some results that demonstrate the usual behaviour of the field dynamics and GW production during preheating.

In Sect. 3.4, I will show that the amplitude of GWs strongly depends on the initial value of the preheating field χ. The main result is presented in Sect. 3.5, where I will demonstrate that the GW background in this model has a scale invariant spectrum of fluctuations, with anisotropies of the order of 1 %. I will discuss how this effect is related to the field dynamics. Finally, I will conclude in 3.6 and give an outlook on future work that could be done in this field.

Note that in this chapter, I will use the Planck mass M_{Pl} instead of the more commonly used reduced Planck mass m_{Pl}, which differ by a factor of $\sqrt{8\pi}$. The simulation code is based on the publicly available ClusterEasy [7], which introduces dimensionless parameters that are rescaled in terms of the Planck mass. To make comparison with the simulation results simpler, all other equations will also be given in terms of M_{Pl}.

3.1 Gravitational Wave Production During Preheating

3.1.1 Studying Preheating Numerically

The analytic study of preheating presented in Sects. 1.4.2 and 1.4.3 is valid up to the point where the system becomes non-linear. This happens when the backreaction of the produced particles becomes large enough to induce correction terms in the inflaton potential and eventually terminate the resonance. Although analytical estimates of when this occurs have been obtained in [8, 9], as in any problem that contains highly non-linear equations, the actual dynamics are best investigated using numerical methods.

In the case of preheating, numerical simulations were first performed by [10], where only a quartic self-coupling of the inflaton was considered. The classical field equation was solved numerically, and the quantum nature of the problem was taken care of by setting up fluctuations in the modes $\phi_{\mathbf{k}}$ as random initial conditions. In [11], interactions with another scalar field were introduced and studied numerically, using lattice field theory simulations where the fields are discretized on a regular lattice.

These early numerical simulations gave new insight into the resonant behaviour for different couplings and the dynamics after the onset of non-linearities.

The non-linear regime of preheating is particularly interesting, as it can lead to the production of a large amount of gravitational waves: due to the amplification of specific momentum modes during preheating, after enough energy has been transferred we are left with large, time-dependent inhomogeneities in the classical field distribution which act as a source for gravitational waves.

Specifically, if a momentum k_* is amplified, this results in field inhomogeneities in configuration space of size $L_* \sim 1/k_*$ which introduce an anisotropic stress term into the stress energy tensor, the transverse-traceless part of which acts as a very efficient source of GWs, see Sect. 1.5.2. Although initially GWs are produced on scales corresponding to the amplified momenta, eventually the inhomogeneous configurations collide and break up into smaller inhomogeneities, which leads to the production of GW on smaller scales $k > k_*$ [12]. After the fields relax and the parametric resonance stops, we are left with a spectrum of GWs which is peaked around k_*, and its shape will carry information about the generation process.

GW production during preheating is a highly non-linear process and therefore needs to be studied numerically. This was first done in [13] and more recently in [4, 14] for the simple chaotic inflation models considered in Sects. 1.4.2 and 1.4.3. Furthermore, numerical simulations of gravitational wave production from hybrid preheating [12] or due to fermions [15, 16] have also been performed.

Unfortunately, if the energy scale of inflation is high, gravitational waves from preheating will peak at high frequencies $f > 1$ MHz today. At the time of production, the causal horizon was much smaller than it is now, and no gravitational waves could have been produced on scales larger than the horizon. Hence, it would be difficult to measure them through the B-mode polarization of the CMB, as unlike the scale-invariant tensor perturbations from inflation, they would only affect the very highest multipoles. Instead, we need to resort to direct detection, however detector technology so far is not sensitive to such high frequencies.

Figure 3.1 shows the sensitivity ranges of the currently proposed/operating detectors, and clearly they do not coincide with the range predicted from preheating (see Sect. 3.1.3 on how to obtain the frequency and energy density of GWs from preheating today). Note that high frequency detectors, capable of measuring signals around 100 MHz, have recently been proposed [17–19]. While their sensitivity is currently too low to detect gravitational waves from preheating, there remains hope that detector technology could evolve to the extent where we are able to directly probe this regime in the future.

3.1.2 Gravitational Waves from Massless Preheating

In the following sections, I will investigate the effect of a light scalar field χ, coupled to the inflaton during preheating, on gravitational wave production. Numerically, it

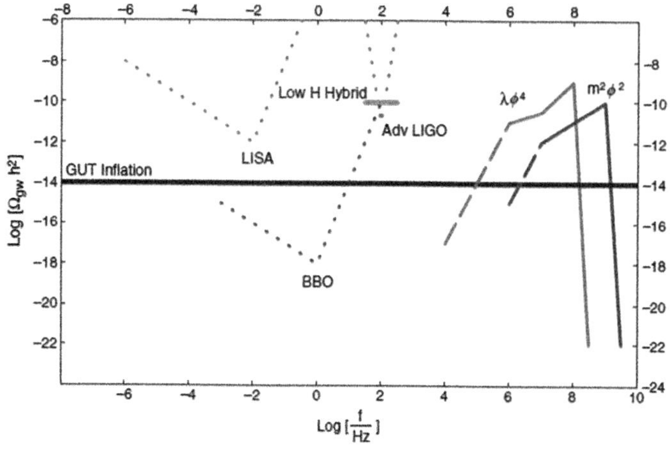

Fig. 3.1 The sensitivity ranges of detectors LIGO, eLISA and BBO, and the ranges at which GW production from preheating peaks for different models. The *straight line* is the scale invariant background from inflation. Figure reproduced from [4]. ©SISSA Medialab Srl. Reproduced by permission of IOP Publishing. All rights reserved

will be easiest to do this for the massless preheating model with a quartic inflaton potential,

$$V(\phi, \chi) = \frac{\lambda}{4}\phi^4 + \frac{1}{2}g^2\phi^2\chi^2 . \tag{3.1}$$

Remember that this model is scale invariant (as the coupling constant is dimensionless, it contains no fixed physical length scale, unlike a model with a mass term $m^2\phi^2$) which resulted in equations of motion that were independent of the scale factor. This makes it particularly convenient for solving numerically on a lattice, as one does not have to take the expansion of the universe into account when performing the simulations and can therefore use a fixed lattice size that will cover the dynamical range at all time steps.

A pure $\lambda\phi^4$ model has been ruled out by the recent Planck data [20]. However, we can modify the potential by including a non-minimal coupling of the inflaton to gravity of the form $\xi\phi^2 R$. For $\xi < 0$, this makes the model viable again as it brings down the tensor to scalar ratio r [21]. If the high value of r suggested by [22] is confirmed by other experiments, this means that even with a weak non-minimal coupling $|\xi| < 0.1$ we can achieve consistency with the data. This term was not included in the simulations as the B-mode discovery was too late to be accounted for, however it should not strongly affect the results, as $|\xi|$ is small and the term will only be significant for large field values of ϕ during inflation.

The Lagrangian for two interacting scalar fields is

$$\mathcal{L} = -\frac{1}{2}\partial_\mu\phi\partial^\mu\phi - \frac{1}{2}\partial_\mu\chi\partial^\mu\chi - V(\phi, \chi) , \tag{3.2}$$

where $V(\phi, \chi)$ is given by Eq. (3.1). The background field evolution can then be written as [using Eq. (1.125)]

$$\ddot{\phi} + 3H\dot{\phi} - \frac{1}{a^2}\nabla^2\phi + (\lambda\phi^2 + g^2\chi^2)\phi = 0\,, \tag{3.3}$$

$$\ddot{\chi} + 3H\dot{\chi} - \frac{1}{a^2}\nabla^2\chi + g^2\phi^2\chi = 0\,. \tag{3.4}$$

The evolution of the background is determined by the Hubble rate. This is given by the Friedmann equation (1.15), where the total energy density is the sum of kinetic, gradient and potential terms (which should be understood as spatially averaged)

$$H^2 = \frac{4\pi}{3M_{\rm Pl}^2}\left[\dot{\phi}^2 + \dot{\chi}^2 + (\nabla\phi)^2 + (\nabla\chi)^2 + 2V(\phi, \chi)\right]. \tag{3.5}$$

Gravitational waves correspond to transverse and traceless tensor perturbations, which I will simply refer to as h_{ij} with $\partial_i h_{ij} = h_{ii} = 0$, dropping the TT superscript. The full spatial metric is therefore given by

$$g_{ij} = a^2(t)(\delta_{ij} + h_{ij})\,. \tag{3.6}$$

These tensor perturbations are sourced by $\Pi_{ij}^{\rm TT}$, the TT part of the anisotropic stress tensor, and their equation of motion in an expanding background is given by Eq. (1.125),

$$\ddot{h}_{ij} + 3H\dot{h}_{ij} - \frac{1}{a^2}\nabla^2 h_{ij} = \frac{16\pi}{M_{\rm Pl}}\Pi_{ij}^{\rm TT}(\phi, \chi)\,. \tag{3.7}$$

We need to determine the anisotropic stress tensor from preheating. We cannot expect the scalar fields to behave like a perfect fluid, as the resonant amplification of momentum bands makes the distribution very inhomogeneous. Therefore, Π_{ij} is not simply a spatial perturbation around a perfect fluid background. However, it is reasonable to define the anisotropic stress tensor as the full energy-momentum tensor minus the isotropic stress, which is given by the background homogeneous pressure [14]:

$$a^2\Pi_{ij} = T_{ij} - \langle P\rangle g_{ij}\,, \tag{3.8}$$

where g_{ij} is the full metric including perturbations, the scale factor a^2 comes from the background FRW metric and we have only kept terms to first order in perturbations. Using Eq. (1.24) for the energy-momentum tensor of both scalars, we find

$$\Pi_{ij} = \frac{1}{a^2}\left[\partial_i\chi\,\partial_j\chi + \partial_i\phi\,\partial_j\phi + g_{ij}(\mathcal{L} - \langle P\rangle)\right]. \tag{3.9}$$

Note that if we do not have a perfect fluid background, there should be an additional term $h_{ij}\left(H^2 + 2\frac{\ddot{a}}{a}\right)$ on the LHS of Eq. (3.7) coming from the perturbed Einstein equations [23]. This term ordinarily cancels with the isotropic pressure perturbation $P\,h_{ij}$ of a perfect fluid [which you can see from the Friedmann equations, Eqs. (1.15) and (1.16)] and therefore does not appear in the equation of motion of tensor perturbations. In this case, where the background is not determined by a perfect fluid, we should include this term when we calculate h_{ij}. However, as gravitational wave production happens on subhorizon scales for which $k \gg aH$, we can ignore the expansion of the universe and the additional term does not need to be taken into account [24].

After applying the TT projection, the term proportional to the metric g_{ij} in (3.9) vanishes: This is because after the TT projection, only the tensor perturbation h_{ij} survives. As the term in brackets is also of order $\mathcal{O}(h)$ [recall Eq. (1.27)], this results in a second order perturbation which can be neglected [12]. Therefore, the tensor modes are simply sourced by the TT projection of the field gradients.

To study fluctuations, we need to write Eq. (3.7) in Fourier space. I will use the following Fourier space convention:

$$\tilde{f}(\mathbf{k}, t) = \int d^3\mathbf{x}\, f(\mathbf{x}, t) e^{-i\mathbf{k}\cdot\mathbf{x}}\,, \tag{3.10}$$

$$f(\mathbf{x}, t) = \int \frac{d^3\mathbf{k}}{(2\pi)^3} \tilde{f}(\mathbf{k}, t) e^{i\mathbf{k}\cdot\mathbf{x}}\,. \tag{3.11}$$

The equation of motion for the fluctuations is given by

$$\ddot{h}_{ij}(t, \mathbf{k}) + 3H\dot{h}_{ij}(t, \mathbf{k}) - \frac{k^2}{a^2} h_{ij}(t, \mathbf{k}) = \frac{16\pi}{M_{\rm Pl}} \Pi^{\rm TT}_{ij}(t, \mathbf{k})\,. \tag{3.12}$$

It is easy to perform the TT projection of the source term in Fourier space by defining a projector

$$\Lambda_{ij,lm}(\hat{k}) = P_{il} P_{jm} - \frac{1}{2} P_{ij} P_{lm}\,, \tag{3.13}$$

$$P_{ij} \equiv \delta_{ij} - k^{-2} k_i k_j\,. \tag{3.14}$$

Using $\Lambda_{ij,lm}$, we can write

$$\Pi^{\rm TT}_{ij}(\mathbf{k}, t) = \Lambda_{ij,lm}(\hat{k}) \int d\mathbf{x}\, e^{-i\mathbf{k}\mathbf{x}} \frac{1}{a^2} \left[\partial_l \chi\, \partial_m \chi + \partial_l \phi\, \partial_m \phi\right](\mathbf{x}, t)\,. \tag{3.15}$$

This projection guarantess that $\Pi^{\rm TT}_{ii}(\mathbf{k}, t) = k_i \Pi^{\rm TT}_{ij}(\mathbf{k}, t) = 0$, $\forall\, \mathbf{k}, t$.

Using a field redefinition to express the wave equation (3.12) as one in flat space which can be solved by a Green function $\mathcal{G}(k, t - t')$ (c.f Sect. 1.5.2), you find that the perturbation h_{ij} has solution [16]

$$h_{ij}(\mathbf{k}, t) = \frac{16\pi}{M_{\rm Pl}^2} \int_{t_i}^{t} dt' \mathcal{G}(k, t - t')\, \Pi_{ij}^{\rm TT}(\mathbf{k}, t')\,, \tag{3.16}$$

where the initial conditions are $h_{ij}(k, t_i) = \dot{h}_{ij}(k, t_i) = 0$. However, I will show in Sect. 3.3.1 that we do not actually need to know the Green function when performing the numerical calculation.

The stress-energy tensor $t_{\mu\nu}$ of gravitational waves, which describes the energy carried by them, is given by Eq. (1.28). The energy density $\rho_{\rm GW} = t_{00}$ can therefore be written as

$$\rho_{\rm GW} = \frac{M_{\rm Pl}^2}{32\pi} \frac{1}{L^3} \int d^3x\, \dot{h}_{ij}(t, \mathbf{x}) \dot{h}_{ij}^*(t, \mathbf{x})\,, \tag{3.17}$$

where I have averaged over the lattice volume $V = L^3$. Writing $\rho_{\rm GW} = \int \frac{d\rho_{\rm GW}}{d\log k} d\log k$, we can define the spectrum of gravitational waves in Fourier space (where the additional factor of $(2\pi)^3$ comes from the Fourier transform):

$$\frac{d\rho_{\rm GW}}{d\log k} \equiv \frac{k^3 M_{\rm Pl}^2}{(4\pi L)^3} \int \frac{d\Omega_k}{4\pi} \dot{h}_{ij}(t, k, \hat{\mathbf{k}}) \dot{h}_{ij}^*(t, k, \hat{\mathbf{k}})\,, \tag{3.18}$$

where $d\Omega_k$ is the solid angle in $\mathbf{k}$ space. Later, I will calculate the total energy density of gravitational waves, normalized to the critical energy density ρ_c,

$$\Omega_{\rm GW}(t) = \frac{1}{\rho_c} \int \left(\frac{d\rho_{\rm GW}}{d\log k}\right) d\log k\,. \tag{3.19}$$

3.1.3 Gravitational Wave Background from Preheating Today

In the simulations I will obtain spectra of gravitational waves, with a specific peak momentum and energy density that can be obtained by integrating over all momenta. However, to predict what the GW background would look like now, we need to relate the frequency and energy to their values today. Due to the weakness of gravity, the waves decouple upon production, so their frequency is simply redshifted,

$$f \equiv \left(\frac{a}{a_0}\right) \frac{k}{2\pi}\,, \tag{3.20}$$

where a and a_0 are the scale factor at the beginning of gravitational wave production and today, respectively, and k is the comoving wave number, related to the physical wave number as $k = k_{\rm phys}(t)a(t)/a$.

We therefore need to find an expression for the ratio of the scale factors, which will depend on two important stages, the end of gravitational wave production and the onset of radiation domination after preheating. Expressing the scale factors in terms of the energy density at these times, you can obtain an expression for the frequency today in terms of parameters defined at the time of preheating [16]:

$$f \approx \left(\frac{a_*}{a_{\rm RD}}\right)^{\frac{1-3w}{4}} \left(\frac{a}{a_*}\right)\left(\frac{k}{\rho_*^{1/4}}\right) \times 5 \cdot 10^{10}\text{Hz}\,, \tag{3.21}$$

where quantities with an asterisk are evaluated at the end of gravitational wave production and $a_{\rm RD}$ is the scale factor at the onset of radiation domination. This is the most general expression, valid for any equation of state $w = P/\rho$ between t_* and $t_{\rm RD}$.

In the case of massless preheating, the background evolves like radiation (see Sect. 1.4.3), which gives $w = 1/3$ and hence the first term of Eq. (3.21) is unity. Furthermore, we can relate the energy density at the end of GW production to its value at the beginning, as during RD $a^4\rho = a_*^4\rho_*$, cancelling the remaining scale factors in (3.21).

The simulations described in the next sections begin at the end of inflation (for which I set $a = 1$), and gravitational wave production starts very soon afterwards. Therefore, we can write the energy density as dominated by the inflaton potential, $\rho = \frac{1}{4}\lambda\phi_{\rm i}^4$, where $\phi_{\rm i}$ is the inflaton value at the beginning of the simulation. The comoving momentum will be defined in units of $\sqrt{\lambda}\phi_{\rm i}$, so I can write the frequency today as

$$f \approx \left(\frac{k}{\rho^{1/4}}\right) \times 5 \cdot 10^{10}\,\text{Hz} \approx \frac{k}{\sqrt{\lambda}\phi_{\rm i}}\lambda^{1/4} \times 7 \cdot 10^{10}\,\text{Hz}\,. \tag{3.22}$$

We will also need to relate the energy density $\Omega_{\rm GW}$ of produced gravitational waves, Eq. (3.19), to its value $\Omega^0_{\rm GW}$ today. Following similar arguments as for the frequency, for massless preheating this is given by [16]:

$$h^2\Omega^0_{\rm GW} = h^2\Omega_{\rm rad}\left(\frac{g_0}{g_*}\right)^{1/3}\Omega_{\rm GW}\,, \tag{3.23}$$

where g_0 and g_* are the number of relativistic degrees of freedom today and during preheating, respectively, and $h^2\Omega_{\rm rad} = 4 \times 10^{-5}$ is the fractional energy density of radiation today. Using $g_*/g_0 \approx 100$, we can therefore rewrite this as

$$h^2\Omega^0_{\rm GW} \approx 9 \times 10^{-6}\,\Omega_{\rm GW}\,. \tag{3.24}$$

As was mentioned in Sect. 1.5.4, GW detectors are sensitive to the amplitude (strain) of the wave and not the energy density. They are related in terms of the frequency as [3]

$$h_{\rm GW}(f) \simeq 1.263 \times 10^{-18}\frac{1\text{Hz}}{f}\sqrt{h^2\Omega^0_{\rm GW}(f)}\,. \tag{3.25}$$

Even for a wave with a large energy density $h^2\Omega^0_{\mathrm{GW}} \approx 10^{-9}$ (see Fig. 3.1), if the frequency is around 10MHz, this would imply a tiny amplitude of $\mathcal{O}(10^{-30})$, which is the reason why GWs from the very early universe are so hard to detect.

3.2 Massless Preheating with a Light Scalar Field

In this section, I will describe the significance of preheating with a light scalar field χ which varies on superhorizon scales. This variation will provide initial conditions for the homogeneous field value χ_{i} to be used in the simulations, and therefore affects the GW production in different preheating volumes.

3.2.1 The Separate Universe Approximation

In Sect. 1.2.5 I have argued that any light field (with a mass less than the inflationary Hubble rate) will acquire a scale invariant spectrum of perturbations from inflation. This is because fluctuations in such a field would freeze out after their comoving modes exit the horizon, just like for the inflaton itself. In contrast, a heavy field is not affected by the damping term due to H and would simply roll down to the bottom of its potential.

In the case of massless preheating, the lightness of the field implies $m_\chi = g\phi < H$. The power spectrum of χ fluctuations is given by the same expression as the inflaton spectrum, Eq. (1.70),

$$\mathcal{P}_\chi \equiv \frac{\partial\langle\chi^2\rangle}{\partial \log k} \simeq \frac{H^2}{4\pi^2}, \tag{3.26}$$

where the definition of the power spectrum as the power per logarithmic k interval, Eq. (1.34), was used. We need to determine for which values of g the field χ is light, such that it satisfies $m_\chi = g\phi < H$, which depends on the value of the inflaton field. At a time N e-foldings before the end of inflation, it is given by $\phi = \sqrt{N/\pi}M_{\mathrm{Pl}}$ [25]. Therefore, χ is light N e-foldings before the end of inflation if

$$\frac{m_\chi^2}{H^2} = \frac{3g^2\phi^2 M_{\mathrm{Pl}}^2}{2\pi\lambda\phi^4} = \frac{3g^2}{2N\lambda} \lesssim 1. \tag{3.27}$$

In order for this to be the case for the largest observable scales, which left the horizon $N \sim 60$ e-foldings before the end of inflation, the couplings must satisfy $g^2/\lambda \lesssim 2N/3 \sim 40$. We want the condition to be satisfied long enough for large scale fluctuations of the field χ to be significantly amplified, as once the Hubble rate falls below m_χ, the field starts oscillating with a decreasing amplitude. In the simulations, the value $g^2/\lambda = 2$ was chosen, which guarantees that χ is light apart from the last few moments of inflation.

The lightness of the field ensures that χ will vary on superhorizon scales and therefore take a different value in different preheating volumes. To accurately model the preheating process, one should consider separate universes, each with a different initial value χ_i. However, we can choose the same initial value ϕ_i for all of them, as this will simply determine at what point in the inflaton's evolution the simulation starts.

Although the initial homogeneous value of χ is many orders of smaller than that of the inflaton field (for χ to be subdominant during inflation) and is often set to zero, it should not be ignored if χ is light as it will provide different initial conditions for the separate universes.

To study the GW background from preheating, I will therefore consider a range of χ_i values (as described in the next section), and perform separate lattice simulations for each of them. The homogeneous field values will be superimposed with subhorizon vacuum fluctuations.

For the choice of couplings $g^2/\lambda = 2$, the homogeneous mode $\kappa = 0$ is within the instability band, see Fig. 1.2. Hence, while initially the linear evolution proceeds very similarly for the different preheating volumes, at the time the dynamics become non-linear the homogeneous mode has been significantly amplified and will have a strong impact on the evolution. Consequently, any quantity that depends on χ_i will vary between different preheating horizon volumes.

What does this imply for the GW background from preheating today? With time, the separate preheating volumes will come into causal contact as the comoving horizon $(aH)^{-1}$ grows, which is about 60 e-folds larger today than at the end of inflation [25]. Therefore, there is a very large number of preheating patches in our current Hubble volume.

Specifically, on Earth we observe GWs originating from a comoving spherical shell of radius $R \sim 1/H_0$, with H_0 the Hubble rate today, and any direction $\hat{n}$ points to a primordial preheating volume at $\mathbf{r} = R\hat{n}$. These regions correspond to a tiny angular size on the sky, much smaller than the 1° angular scales which correspond to the size of the horizon at last scattering. Although this means that we cannot distinguish between individual preheating volumes, the GW energy density, which is a function of position, $\Omega_{\rm GW}(\hat{n}) = \Omega_{\rm GW}\left[\chi_i(R\hat{n})\right]$, can vary on cosmological scales. Hence, we expect the GW background from preheating with a light scalar field to be anisotropic.

3.2.1.1 Impact of Light χ on Curvature Perturbations

Before describing which range of χ_i values we should consider, I want to comment on the effect of preheating with a light scalar χ on the curvature perturbation.

The field fluctuations $\chi_{\mathbf{k}}$ that are amplified during preheating represent an isocurvature perturbation, i.e. they do not vanish on spatially flat hypersurfaces. In [26] it was shown that such a contribution could have an effect on the curvature perturbation ζ, defined in Eq. (1.71). This can easily be seen within the separate universe approximation. The difference in the evolution between different FRW volumes affects ζ as $\delta\zeta = \delta N$ [26], where $N = \ln a$. As the evolution of each volume will depend on the initial value $\chi_{\rm i}$, we can therefore expect a contribution towards the total curvature perturbation from preheating.

This was first investigated by [27, 28], where a random contribution to δN was observed which would manifest itself as white noise in the data. However, these references neglected to include inhomogeneous modes in their simulations which were taken into account by [1, 29, 30]. In particular, the more accurate simulations in [1] demonstrated a highly non-Gaussian structure on top of a random background: for certain periodically spaced values of $\chi_{\rm i}$, they observed spikes in δN which could have a measurable effect on ζ that would contribute to cold spots in the CMB temperature. This occurred for values of $\chi_{\rm i}$ that resulted in a very large amplification of the homogeneous mode of χ, much larger than the initial inflaton amplitude.

Clearly, it would be interesting to study the correlation between the curvature perturbation and the GWs produced during preheating. However, the numerical algorithm I used, see Sect. 3.3.1, was not accurate enough to calculate the curvature perturbation, which is related to changes in the scale factor of order $\mathcal{O}(10^{-5})$ [25]. I will briefly comment on how the same field dynamics that lead to spikes might affect the GW amplitude in Sect. 3.5.3.

3.2.2 *Varying $\chi_{\rm i}$ During Preheating*

To calculate the GW background from preheating, we first need to determine what range of $\chi_{\rm i}$ values we can expect the GW background from preheating to have originated from. Since $\chi_{\rm i}$ is a Gaussian random field with a scale-invariant spectrum (3.26), it will have a non-zero average value in any given volume, even in the comoving volume that corresponds to the currently observable universe. This is because fluctuations that are much larger than the current horizon have been amplified by inflation, as long as it lasted longer than the minimum 60 e-folds, which is likely.

The total range of amplified, comoving wavelengths extends from the Hubble length at the end of inflation, $k \sim H_*$ (well inside the horizon today), to the Hubble length at the start of inflation, which probably corresponds to a superhorizon scale much larger than our current horizon.

From the observational point of view, the wavelengths that are currently inside the horizon, $k \gtrsim a_0 H_0$, appear as inhomogeneous fluctuations, or anisotropies on the sky. The variance σ_χ^2 of these fluctuations can be computed from the power spectrum (3.26),

$$\sigma_\chi^2 = \int_{a_0 H_0}^{H_*} \frac{dk}{k} \mathcal{P}_\chi = \frac{H_*^2}{4\pi^2} \ln \frac{H_*}{a_0 H_0} = \frac{H_*^2}{4\pi^2} N_* , \tag{3.28}$$

where the ratio of the comoving Hubble horizon today, $(a_0 H_0)^{-1}$, to the one at the end of inflation, H_*^{-1} (where I set $a_* = 1$), is given by $N_* \sim 60$, the number of e-folds of inflation after the largest observable scales left the inflationary Hubble radius.

If inflation lasted longer than $N_* \sim 60$ e-folds, even larger scales were amplified and χ will have varied on scales that are superhorizon now. The actual mean value $\overline{\chi}_{\rm i}$ across the universe would be a particular realization drawn from a Gaussian distribution with variance

$$\langle \overline{\chi}_{\rm i}^2 \rangle = \int_{(aH)_{\rm start}}^{a_0 H_0} \frac{dk}{k} \mathcal{P}_\chi = \int_{(aH)_{\rm start}}^{H_*} \frac{dk}{k} \mathcal{P}_\chi - \int_{a_0 H_0}^{H_*} \frac{dk}{k} \mathcal{P}_\chi$$
$$= \frac{H_*^2}{4\pi^2} (N_{\rm tot} - N_*) \,, \tag{3.29}$$

where $N_{\rm tot} = \ln(1/a_{\rm start})$ (remember $H \approx H_*$ throughout inflation) is the total number of e-foldings of inflation. A typical average field value across a volume as large as our observable universe is then

$$\overline{\chi}_{\rm i} \sim \frac{H_*}{2\pi} \sqrt{(N_{\rm tot} - N_*)} \,. \tag{3.30}$$

Since the value of $N_{\rm tot}$ is unknown, I will consider the actual realization of $\overline{\chi}_{\rm i}$ within our observable patch as a free parameter, simply restricted to $\overline{\chi}_{\rm i} > H_*/2\pi$. In the simulations, I will study the dependence of $\Omega_{\rm GW}$ on different values of $\chi_{\rm i}$, drawn from a Gaussian distribution with the variance given in Eq. (3.28), and centred around a mean value $\overline{\chi}_{\rm i}$ of order of Eq. (3.30).

3.3 Numerical Simulations

3.3.1 Numerical Algorithm

To study the GW production for different initial values $\chi_{\rm i}$, I performed simulations on a 3d lattice with periodic boundary conditions, populated with the fields χ, ϕ and the six tensor perturbation components h_{ij}. The code I used is based on the publicly available ClusterEasy [7], an MPI/C++ package performing lattice simulation of interacting scalar fields in an expanding universe.

The numerical algorithm used to solve the differential equations is a second-order leapfrog integrator where field values and their derivatives are stored at different times. Although this method is not as accurate as more popular fourth order Runge-Kutta methods, its advantages are its simplicity and speed. In the case under consideration, where we are interested in gravitational wave production, it is not necessary to have an extremely accurate integrator. It can be checked that the solution is stable by observing the evolution of quantities like the total energy (which should be conserved), and by ensuring that the chosen time step does not affect the results. The difference in GW energy which will be a large $\mathcal{O}(1)$ effect, as opposed to measuring the difference in scale factor within the δN formalism, which is $\mathcal{O}(10^{-5})$. Therefore, to calculate curvature perturbations from preheating, a more accurate integrator is needed, see [1].

To study the field evolution we need to solve discretized versions of the field equations for the scalars, Eqs. (3.3) and (3.4), and the Friedmann equation (3.5). The evolution of the scale factor is solely determined by the scalar fields, and I checked that it indeed evolves as $a \propto t^{1/2}$, as if dominated by radiation. I will assume that each separate preheating volume can be described by an FRW background metric. This is justified as long as the lattice volume does not strongly exceed the comoving horizon at the time which determines the spatial extent of causally connected regions. We know that the variation in scale factor between different volumes is of order 10^{-5} [25], and should therefore not have a strong effect on the dynamics of the scalar fields.

The evolution of the tensor perturbations, which determines the GW spectrum, is given by Eq. (3.7). I chose not to include backreaction from the tensor perturbations into the scalar field equations, as these were shown to be negligible for GW production during preheating in [12]. We can see that this should be the case, as we know from the Lagrangian (3.2), which contains the metric $g_{\mu\nu}$ in the derivative terms, that they will appear as $h_{ij}\partial_i\chi\partial_j\chi$ in the equation of motion for χ. As h_{ij} is a small perturbation, this is clearly negligible compared to the usual derivative term, and can therefore be ignored.

To compute the spectrum (3.18), in principle, for each time step, we need to perform the TT projection in Eq. (3.7), then go to Fourier space to solve the equation, and finally transform back to coordinate space. Both the TT projection and the Fourier transforms are non-local operations and therefore computationally very costly. To avoid this, I followed the method introduced in [12], which makes use of the TT projector $\Lambda_{ij,lm}$ defined in Eq. (3.13). We saw there that the solution of Eq. (3.7) can formally be written in terms of a Green function, Eq. (3.16). This can be re-written in terms of a function $u_{ij}(\mathbf{k}, t)$, related to the tensor perturbation by the projection operator,

$$\dot{h}_{ij}(\mathbf{k}, t) = \Lambda_{ij,lm}(\hat{k})\dot{u}_{lm}(\mathbf{k}, t)\,. \tag{3.31}$$

The solution of $\dot{u}_{ij}$ is given by

$$\dot{u}_{ij}(\mathbf{k}, t) \equiv \frac{16\pi}{M_{\mathrm{Pl}}^2}\int_{t_i}^{t} dt'\dot{\mathcal{G}}(k, t - t')\,\Pi_{ij}^{\mathrm{eff}}(\mathbf{k}, t')\,. \tag{3.32}$$

I have introduced an effective anisotropic stress $\Pi_{ij}^{\mathrm{eff}}(\mathbf{k}, t)$, which is the Fourier space version of the unprojected source term

$$\Pi_{ij}^{\mathrm{eff}}(\mathbf{x}, t) \equiv \frac{1}{a^2}\left[\partial_i\chi\partial_j\chi + \partial_i\phi\partial_j\phi\right](\mathbf{x}, t)\,, \tag{3.33}$$

c.f. Eq. (3.15). Having rephrased the equations in this manner enables us to avoid having to perform the TT projection explicitly. Instead, during the simulation, at each time step I solve the equation of motion for u_{ij} in configuration space,

$$\ddot{u}_{ij} + 3H\dot{u}_{ij} - \frac{1}{a^2}\nabla^2 u_{ij} = \frac{16\pi}{M_{\rm Pl}^2}\Pi_{ij}^{\rm eff}(\phi, \chi)\,. \tag{3.34}$$

Only at the times when we want to compute the GW spectrum (3.18), which is determined by the time derivatives $\dot{h}_{ij}$, we Fourier transform $\dot{u}_{ij}(\mathbf{x}, t)$ to $\dot{u}_{ij}(\mathbf{k}, t)$, and recover the real GW degrees of freedom $\dot{h}_{ij}(\mathbf{k}, t)$ by means of the projection in Eq. (3.31).

To calculate the GW spectrum from the lattice simulation, we need to define a discretized version of Eq. (3.17):

$$\rho_{\rm GW} = \frac{M_{\rm Pl}^2}{32\pi}\frac{1}{N^3}\sum_{\mathbf{n}}\dot{h}_{ij}(t, \mathbf{n})\dot{h}_{ij}^*(t, \mathbf{n})\,, \tag{3.35}$$

where I used $L^3 = (N\delta x)^3$, N being the number of lattice points per dimension and $\delta x = L/N$ the lattice spacing, and introduced the discrete position vector $\mathbf{n} = (n_1, n_2, n_3)$ where $n_i = 0, 1, \ldots, N-1$. The discrete Fourier transform is defined by

$$f(\mathbf{n}) = \frac{1}{N^3}\sum_{\mathbf{n}} e^{-\frac{2\pi i}{N}\tilde{\mathbf{n}}\cdot\mathbf{n}}\tilde{f}(\tilde{\mathbf{n}})\,, \tag{3.36}$$

where $\tilde{\mathbf{n}}$ is the discrete momentum vector with integer entries $\tilde{n}_i = -\frac{N}{2} + 1, \ldots, 0, \ldots, \frac{N}{2}$. Using the discrete delta function $\sum_{\mathbf{n}} e^{-\frac{2\pi i}{N}(\tilde{\mathbf{n}}-\tilde{\mathbf{n}}')\mathbf{n}} = N^3\delta(\tilde{\mathbf{n}} - \tilde{\mathbf{n}}')$, we can obtain Eq. (3.35) in Fourier space:

$$\rho_{\rm GW} = \frac{M_{\rm Pl}^2}{32\pi}\frac{1}{N^6}\sum_{\tilde{\mathbf{n}}}\dot{h}_{ij}(t, \tilde{\mathbf{n}})\dot{h}_{ij}^*(t, \tilde{\mathbf{n}})\,. \tag{3.37}$$

To find a simple version of the discretized GW spectrum, it will be necessary to bin the momentum space lattice into spherical layers of radius $|\tilde{\mathbf{n}}|$ and width 1, where $|\tilde{\mathbf{n}}|$ takes integer values between 0, ..., $\sqrt{3}N/2$, and the largest radius corresponds to the absolute value of the momentum vector $\left(\frac{N}{2}, \frac{N}{2}, \frac{N}{2}\right)$. Following the steps outlined in [31], we obtain

$$\rho_{\rm GW} = \sum_{|\tilde{\mathbf{n}}|}\left[\frac{dx^6 M_{\rm Pl}^2}{(4\pi)^3 L^3}\, k^3(|\tilde{\mathbf{n}}|)\left\langle\dot{h}_{ij}(t, |\tilde{\mathbf{n}}|)\dot{h}_{ij}^*(t, |\tilde{\mathbf{n}}|)\right\rangle_{R(\tilde{\mathbf{n}})}\right]\Delta\log k\,, \tag{3.38}$$

where we average over all discrete momenta in a shell $R(\tilde{\mathbf{n}}) = \{\tilde{\mathbf{n}}' \,|\, |\tilde{\mathbf{n}}| \le |\tilde{\mathbf{n}}'| \le |\tilde{\mathbf{n}}| + 1\}$, $k(|\tilde{\mathbf{n}}|) = |\tilde{\mathbf{n}}|\delta k$, $\Delta\log k = \frac{1}{k}\delta k$ and the reciprocal lattice spacing is $\delta k =$

$k_{\rm IR} = 2\pi/L$. The reciprocal lattice spacing corresponds to the smallest infrared momentum, or largest wavelength, that fits into the lattice. The term in square brackets in Eq. (3.38) gives the spectrum for each discrete momentum $k(|\tilde{\mathbf{n}}|) = |\tilde{\mathbf{n}}|\delta k$ and is calculated during the simulation. I will later plot spectra that have been normalized by the critical energy density, which is just determined by the total energy density of the scalar fields.

While the binning is necessary to obtain the power spectrum, to get an accurate measure of the GW energy density $\rho_{\rm GW}$, it is better not to evaluate it using (3.38), but to calculate it in the Cartesian way, Eq. (3.37). This gives more accurate results as it does not assume that points in the same shell at different lattice sites correspond to the same momentum. In the following, the relative, total GW energy density $\Omega_{\rm GW}$ was always calculated using the Cartesian approach.

Note that there are several ways of defining a discretized version of the projection operator $\Lambda_{ij,lm}$ in Eq. (3.31) on a lattice, which depend on the discretization schemes for lattice derivatives. The different projections were analysed in detail in [31]. In the simulations, I used a real projector based on a neutral derivative scheme,

$$[\nabla_i f]\,(\mathbf{n}) \equiv \frac{f(\mathbf{n}+\hat{\imath}\delta x) - f(\mathbf{n}-\hat{\imath}\delta x)}{2\delta x}\,, \tag{3.39}$$

where $\hat{\imath}$ is the unit vector in the i direction. Transforming to Fourier space, where $\widetilde{[\nabla_i f]}(\tilde{\mathbf{n}}) \equiv -i\mathbf{k}_{\rm eff}(\tilde{\mathbf{n}})\tilde{f}(\tilde{\mathbf{n}})$, we can obtain the effective momentum corresponding to the neutral derivative [31],

$$k_{{\rm eff},i} = \frac{\sin(2\pi\tilde{n}_i/N)}{\delta x}\,. \tag{3.40}$$

The discretized projector in Eq. (3.14) for neutral derivatives is therefore given by

$$P_{ij}(\tilde{\mathbf{n}}) = \delta_{ij} - \frac{k_{{\rm eff},i}k_{{\rm eff},j}}{(k_{\rm eff})^2}\,. \tag{3.41}$$

You can check that this definition ensures the transversality and tracelessness of the tensor perturbations when acting on them with $\Lambda_{ij,lm}$. Although only the neutral projector was used for the simulations presented below, I made sure that the results were not affected by the choice of projector.

3.3.2 *Choosing the Numerical Parameters*

To carry out the simulations, it is necessary to specify a number of numerical parameters. Particularly, we need to specify a lattice size L (where the lattice volume is given by L^3) and the number of lattice points N.

The lattice volume is a very important quantity. It should not be significantly bigger than the Hubble horizon at the time of preheating, as otherwise the assumption of a uniform FRW background breaks down. More importantly, the lattice size determines the infrared momentum cutoff $k_{\rm IR} = 2\pi/L$, which corresponds to the largest wavelength that can fit into the simulation box. As the value $g^2/\lambda = 2$ is used, we will need good IR coverage as long wavelength modes are amplified most strongly in this model. However, due to causality, GW modes will not be produced on scales larger than the horizon volume, and there will be a peak scale which depends on the model parameters. Let me give an order of magnitude estimate of this value.

During massless preheating, the width of each amplified (dimensionless) momentum band is given by [9]

$$\Delta\kappa \lesssim \frac{1}{\sqrt{\pi}}\left(\frac{g^2}{\lambda}\right)^{1/4} \approx 0.67\,. \tag{3.42}$$

where $\kappa = k/\sqrt{\lambda}\varphi_{\rm i}$, see Sect. 1.4.3, and I have used $g^2/\lambda = 2$. This is an analytical estimate, valid for values $g^2/\lambda \gtrsim 1$, which becomes more accurate for larger values of g^2/λ. For $g^2/\lambda = 2$, the smallest resonant mode is given by $\kappa = 0$, so Eq. (3.42) gives the largest momentum value that is amplified by the resonance. From the numerical solution in Fig. 1.2, you can see that for $g^2/\lambda = 2$, the principal resonant band seems to be bounded by $\Delta\kappa^2 \lesssim 0.3$, and thus the actual value width of the resonance band is closer to $\Delta\kappa \lesssim 0.55$.

Although the field fluctuations are amplified most strongly the smaller the value of κ, the spectrum of fluctuations, which goes as $k^2|\chi_k|^2 \sim k^2 e^{2\mu(k)}$, will peak at some intermediate scale κ_* between 0 and $\Delta\kappa$, typically a fraction of $\Delta\kappa$, $\kappa_* \sim \mathcal{O}(0.1)$. The fluctuations in the inflaton, on the other hand, depend on the resonance for $g^2/\lambda = 3$, which does not amplify any large wavelengths, and will not affect our choice of L.

The source of GWs, formed by products of fields, inherits the peak scale $\kappa_* \sim \mathcal{O}(0.1)$ of the field spectrum, as can be seen from Eq. (3.15):

$$\Pi_{ij}^{\rm TT}(\mathbf{k}, t) = \Lambda_{ij,lm}(\hat{\mathbf{k}})\frac{1}{a^2}\int d\mathbf{q}\; q_l\, q_m\, \chi_q(t)\, \chi^*_{|\mathbf{k}-\mathbf{q}|}(t)\,. \tag{3.43}$$

As the tensor fluctuations are directly related to the anisotropic stress tensor through Eq. (3.16), this peak scale will also translate to the spectrum of gravitational waves $\frac{d\rho_{\rm GW}}{d\log k}(k, t)$ via Eq. (3.18).

To facilitate comparison with the dimensionless momentum κ, we define the comoving lattice size in the same units, $\widetilde{L} \equiv \sqrt{\lambda}\phi_{\rm i} L$, where L is the physical lattice size. We will see later that the value $\widetilde{L} = 80$, which corresponds to $\kappa_{\rm IR} \approx 0.08$, is sufficient to capture the peak scale.

Note that the comoving Hubble volume at the beginning of our simulation, which is determined by the Hubble rate $H_{\rm i}^2 = \frac{8\pi}{12M_{\rm Pl}^2}\lambda\phi_{\rm i}$, is $1/a\widetilde{H} \approx 2$ in dimensionless program units (where I used the value of $\phi_{\rm i}$ given below). This is quite a bit smaller than $\widetilde{L} = 80$, however the Hubble volume grows to scales larger than the horizon volume before the system becomes non-linear.

While a large enough lattice spacing ensures good IR coverage, we also need to have sufficient UV coverage, which improves with the number of lattice points N, which determines the lattice spacing as $\delta x = L/N$. This needs to be smaller than any relevant length scale in the problem, which in our case is determined by the inverse of the effective mass of the inflaton, $m_\phi^2 = \lambda\phi^2$. As all length scales are rescaled by the mass $\sqrt{\lambda}\phi_{\rm i}$, this implies that in program units we simply need to satisfy $\delta\widetilde{x} < 1$. For a lattice volume of $\widetilde{L} = 80$, the choice $N = 512$ therefore led to a sufficiently small $\delta\widetilde{x} \approx 0.16$. For stability, the program requires [7] the time step to satisfy $\delta t < \delta x/\sqrt{3} \approx 0.1$, and a value $\delta t = 0.01$ was used in the simulations. I checked that these choices led to stable, trustworthy results by ensuring that the total energy in the simulation box was conserved throughout the simulation to high accuracy.

The value of the inflaton self-coupling was set to $\lambda = 9 \times 10^{-14}$, which is the value that is required for consistency with WMAP data [32]. The only parameters left to fix are the initial conditions for the fields.

At the start of every simulation, the scale factor was set to $a_{\rm i} = 1$, and the initial amplitude of the homogeneous inflaton to $\phi_{\rm i} = 0.342 M_{\rm Pl}$, corresponding to the value for which $\dot{\phi}_{\rm i} = -H_*\phi_{\rm i}$ in the slow roll regime. Note that in the simulations, all fields have been rescaled by the scale factor a and are given in units of $\phi_{\rm i}$.

The initial background value $\chi_{\rm i}$ was chosen as described in Sect. 3.2.2. From Eq. (3.5), the Hubble rate at the end of inflation (when the potential term dominates) is given by

$$H_*^2 \approx \frac{8\pi\lambda\phi_{\rm i}^4}{12M_{\rm Pl}^2} \simeq 2.6 \times 10^{-15} M_{\rm Pl}^2. \tag{3.44}$$

Using $N_* \sim 60$ and Eq. (3.28), the variance of $\chi_{\rm i}$ across the observable universe is then $\sigma_\chi^2 \simeq 4 \times 10^{-15} M_{\rm Pl}^2$. Taking $(N_{\rm tot} - N_*) \sim 100$, the mean value of $\chi_{\rm i}$ in Eq. (3.30) is of order $\overline{\chi}_{\rm i} \sim 10^{-7} M_{\rm Pl}$. In the simulations I made the specific choice $\overline{\chi}_{\rm i} = 3.42 \times 10^{-7} M_{\rm Pl}$. However, using the Monte Carlo reweighting method explained in Sect. 3.5.1, the results could be extrapolated to other, neighbouring values of $\chi_{\rm i}$.

To be able to model the parametric resonance, we also need to set up fluctuations ϕ_k and χ_k. These are supposed to mimic quantum fluctuations on scales which are subhorizon after inflation. I followed the approach of [10]: Consider each mode (ϕ_k, χ_k) as given by a complex number $|f_k|e^{+i\varphi_k}$. The phases φ_k are randomly picked from a uniform distribution between $[0, 2\pi)$, while the amplitudes are set according to a Rayleigh distribution with variance

$$\langle |f_k|^2 \rangle = \frac{1}{2a^2\omega_k}, \quad \omega_k \equiv \sqrt{k^2 + m_f^2}, \tag{3.45}$$

where the effective masses are $m_\phi^2 \equiv 3\lambda\phi_i^2 + g^2\chi_i^2$ and $m_\chi^2 \equiv g^2\phi_i^2$. Hence, more massive fields have smaller vacuum fluctuations and the amplitude is smaller for higher momentum modes, which is physically sensible. We should not populate Fourier modes up to arbitrary large momenta, but introduce a cutoff which needs be larger than the peak of the GW spectrum. In the simulations, I used the value $\kappa_* = 2$, but checked that the choice of cutoff did not affect the results.

3.3.3 Simulating Gravitational Wave Production

Having specified all of the numerical parameters, we are now ready to look at the results of the simulations.

Fig. 3.2 shows the evolution of the fields and their variances, and confirms the usual behaviour of parametric resonance as described in 1.4.3. Initially, the amplitude of ϕ is much larger than that of χ, but the oscillations of the former induce a resonant growth of the χ fluctuations. This is shown very clearly by the variance term $\langle\chi^2\rangle$, which grows exponentially fast from from $\tau = 0$ to $\tau = 70$, where τ is the rescaled conformal time, $d\tau = (\sqrt{\lambda}\phi_i/a)dt$. The variance in ϕ also grows due to its self-interactions and coupling to χ, but its growth only starts at around $\tau = 40$, once $\langle\chi^2\rangle$ has already been amplified by around six orders of magnitude. The energy transferred from ϕ to χ is significant, so the (mean) amplitude of χ eventually reaches that of ϕ, at about $\tau = 70$, and the system becomes non-linear. At this point, backreaction

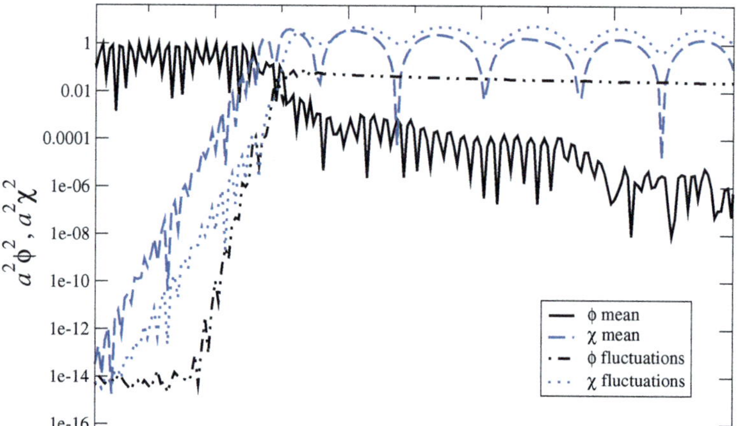

Fig. 3.2 Evolution of the mean field amplitudes squared, $a^2\langle\phi\rangle^2$, $a^2\langle\chi\rangle^2$ and of their variances, $a^2(\langle\phi^2\rangle - \langle\phi\rangle^2)$, $a^2(\langle\chi^2\rangle - \langle\chi\rangle^2)$. The exponential growth of fluctuations due to parametric resonance can be clearly appreciated

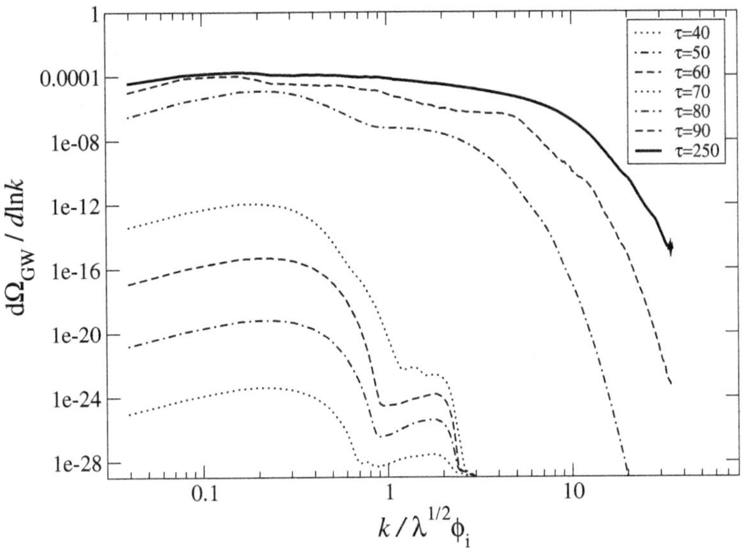

Fig. 3.3 Typical GW spectra from massless preheating with $g^2/\lambda = 2$, shown at different time steps as the amplitude grows. The highest *curve* (continuous line) corresponds to the final time step of our simulation $\tau = 250$, when the amplitude saturates. The peak of the spectrum is at $\kappa_* \sim \mathcal{O}(0.1)$. The production of GWs increases significantly between $\tau = 70$ and $\tau = 80$, when the system becomes non-linear and there is a transfer of power into smaller scales (higher momenta)

from the produced field fluctuations becomes important, reducing the amplitude of the inflaton and terminating the resonance.

While GW production starts as soon as the first field inhomogeneities are introduced due to the exponential growth of the χ fluctuations, it only becomes significant when the dynamics are non-linear. Figure 3.3 shows the GW spectrum plotted at different time steps. The amplified momentum range extends over more than two orders of magnitude, and the spectrum falls off in the UV which shows that it is not dominated by lattice artefacts.

During the linear evolution up to $\tau = 70$, the spectrum peaks at a scale $\kappa_* \approx 0.25$, which is of $\mathcal{O}(0.1)$ as expected from the considerations in Sect. 3.3.2. During the subsequent stage of non-linear evolution, from $\tau = 70$ until $\tau = 100$, the field gradients become much larger, and consequently GWs are being produced with larger intensity. Due to rescattering [9], power is transferred to higher momentum modes. The GW production reaches an end at around $\tau = 150$, however the amplitude is not constant but oscillates slightly as the system enters into a turbulent regime before equilibrating [12]. To obtain the final GW spectrum, it was therefore necessary to average over a few oscillations. This ensured that the value of the calculated gravitational wave energy density Ω_{GW} is trustworthy.

The oscillations of the total GW energy density, obtained by summing over all lattice momenta, is shown in Fig. 3.4. The plot shows the evolution of Ω_{GW} for two

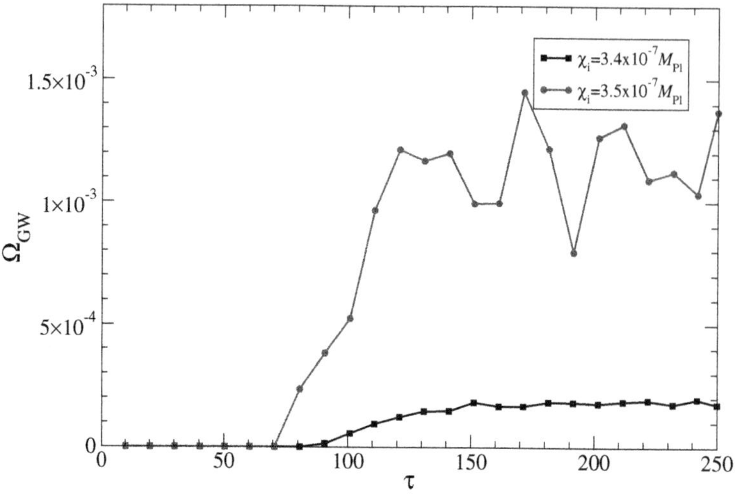

Fig. 3.4 The total energy density of gravitational waves as a function of rescaled conformal time τ for two different initial field values $\chi_{\rm i}$

values of $\chi_{\rm i}$. We can already see that the energy in both cases is very different, and I will explore this in more detail in the next section. The oscillations in the GW energy for $\chi_{\rm i} = 3.5 \times 10^{-7} M_{\rm Pl}$ are quite large. However, this value is unusual in the sense that it leads to an atypically large GW energy, as discussed more below. For most values of $\chi_{\rm i}$, the magnitude of oscillations is of the order of the lower curve in Fig. 3.4.

3.4 The Impact of $\chi_{\rm i}$ on Gravitational Wave Production

The final amplitude of the GW spectrum for the values $\chi_{\rm i} = 3.5 \times 10^{-7} M_{\rm Pl}$ (upper, blue curves) and $\chi_{\rm i} = 3.4 \times 10^{-7} M_{\rm Pl}$ (lower, black curves), chosen for purposes of illustration, is shown in Fig. 3.5. The spectra were obtained from an average over several time oscillations, as shown by the error bars. Before investigating the difference between the two initial values, let me comment on the different types of curves in Fig. 3.5, corresponding to different lattice volumes.

The dashed lines correspond to the fiducial choice of lattice size and number of points per dimension $(\widetilde{L}, N) = (80, 512)$, whereas the solid lines correspond to $(\widetilde{L}, N) = (160, 1024)$, ensuring the same UV coverage. For $\widetilde{L} = 160$, one can clearly see a large drop in the IR, which shows that very long wavelength modes are not excited, as expected from causality.

The runs with $(\widetilde{L}, N) = (160, 1024)$ were computationally too expensive for my purposes: as I want to calculate the GW background on cosmological scales, it was necessary to perform several hundreds of simulations to get a statistical measure of

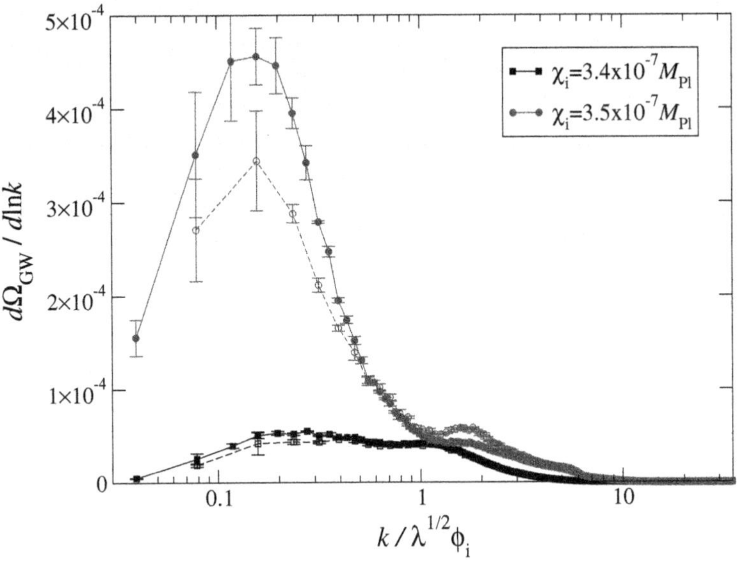

Fig. 3.5 Final spectrum of GWs for $\chi_i = 3.4 \times 10^{-7} M_{Pl}$ (*upper, blue curves*) and $\chi_i = 3.5 \times 10^{-7} M_{Pl}$ (*lower, black curves*), averaged over time oscillations. The error bars show the variation due to this averaging. The *solid curves* are for $\tilde{L} = 160$, $N = 1024$, and the *dashed curves* for $\tilde{L} = 80$, $N = 512$. The area underneath corresponds to the total fractional GW energy density within a preheating Hubble domain

the anisotropy. However, I chose to run a few simulations with such a large lattice volume to show that the spectra for $N = 512$ and $N = 1024$ are comparable. For the majority of initial values, the total integrated GW amplitude from both cases agrees to better than 1 %.

For the upper curves in Fig. 3.5, which have a very high amplitude (which is actually one of the largest achieved values in the simulation, $\Omega_{GW} \simeq 1.2 \times 10^{-3}$, see Fig. 3.8), the difference is a lot larger, around 15 %. However, we can see that the higher resolution case $(\tilde{L}, N) = (160, 1024)$ leads to an even larger difference between the two different initial values χ_i. Therefore, the effect I want to demonstrate, which is the strong dependence of GW amplitude on the initial value, would clearly persist (and even be enhanced) if even better lattice coverage was used.

These considerations show that the fiducial case $(\tilde{L}, N) = (80, 512)$, which is used systematically in Sect. 3.5, is not dominated by lattice artefacts, and can therefore be trusted. For lattices with $N = 256$, independently of the volume $\tilde{L}$, it was not possible to capture both the IR and UV behaviour sufficiently well at the same time. Runs with $(\tilde{L}, N) = (> 80, 512)$ improved the IR coverage but would require to upgrade to $N = 1024$ to keep sufficient UV coverage, which, as mentioned before, was too costly computationally.

The choice $(\widetilde{L}, N) = (80, 512)$ therefore turned out to be the optimal one, representing a good compromise between a sufficiently large dynamical range, and low enough memory usage and shorter duration of the runs.

Let me now comment on the difference between the two spectra in Fig. 3.5. While during the early stages of the simulations the GW spectra evolve in the same way, the homogeneous field value χ strongly affects the production at the time the system becomes non-linear. The peaks of the spectra are located at the same scale $\kappa_* \simeq 0.2$, as this is simply related to the value of the resonant momentum, determined by the choice $g^2/\lambda = 2$. However, the peak GW amplitude is very different (by about a factor of four), even though the initial values χ_i are very similar.

As Fig. 3.5 shows a log-linear plot, the area underneath the curves corresponds to the total fractional GW energy density within a preheating volume, which is clearly also going to differ significantly between the two cases. This $\mathcal{O}(1)$ effect is much larger than what could be naively expected.

One might worry that the difference in amplitude demonstrated in Fig. 3.5 could be just a statistical effect related to the initial conditions for the UV modes. However, I checked that the difference in GW amplitude due to statistical fluctuations (by choosing different random seeds) is much smaller than the difference in amplitude between different initial values. When I present the variation of GW energy for a large range of χ_i later, the error bars due to the statistical fluctuation are included, see Fig. 3.8. Clearly, it is very small compared to the actual effect. The final discrepancy in amplitude of the GW spectra must therefore arise because of the different behaviour of the fields sourcing the GWs, which is ultimately related to the initial amplitude χ_i.

Because the GWs are sourced by field gradients, the homogeneous component has no effect until the evolution becomes nonlinear. At this point, the energy in the homogeneous mode is redistributed among other momenta. Different values of χ_i will therefore create a different outcome in the spatial distribution of χ.

In Fig. 3.6, I show a time sequence of 2d snapshots of the 3d spatial distribution of the field χ. Compared to Fig. 3.5, they were obtained for two different initial values $\chi_\mathrm{i} = 3.4 \times 10^{-8} M_\mathrm{Pl}$ (left panels) and $\chi_\mathrm{i} = 1.0 \times 10^{-8} M_\mathrm{Pl}$ (right panels), but they can be used to illustrate the physical reason for the difference in GW amplitude. The GW energy density varies significantly between the simulations in Fig. 3.6, $\Omega_\mathrm{GW} = 1.1 \times 10^{-3}$ and $\Omega_\mathrm{GW} = 5.6 \times 10^{-4}$.

The snapshots are taken at times during the non-linear evolution of the fields, in $\Delta\tau = 2$ intervals between $\tau = 73$ and $\tau = 79$, just when the GW production is strongest. Fig. 3.6 demonstrates very clearly that there is a correlation between the gradients of χ and the amplitude of the produced GWs: for $\chi_\mathrm{i} = 3.4 \times 10^{-8} M_\mathrm{Pl}$, the gradients and, consequently the GW amplitude, are higher than for $\chi_\mathrm{i} = 1.0 \times 10^{-8} M_\mathrm{Pl}$. The physical reason for the sensitive dependence of the gradients of χ on the initial value χ_i will be investigated in more detail in Sect. 3.5.3.

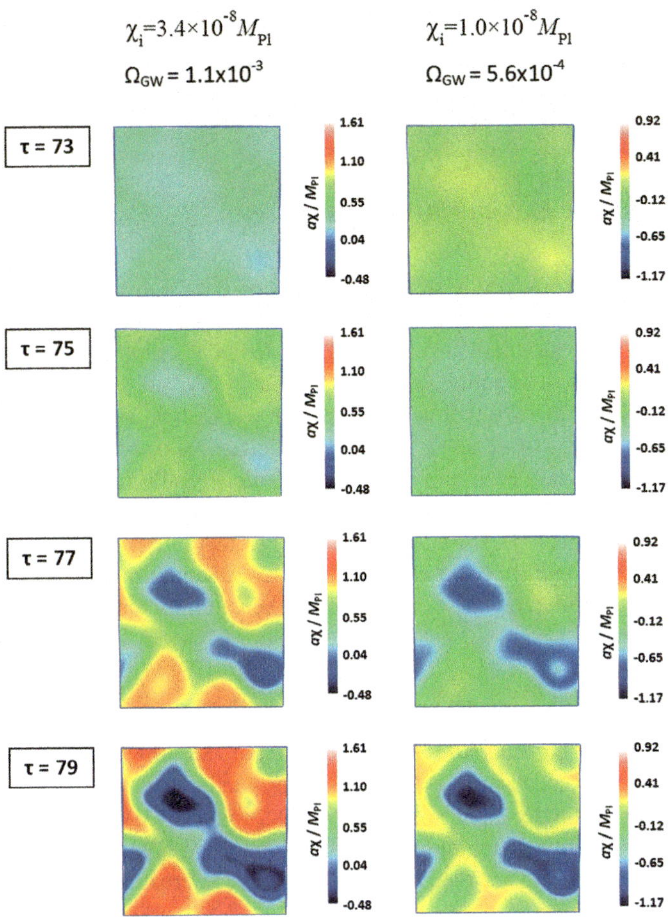

Fig. 3.6 2d snapshots of the 3d distribution of χ at different times of the evolution during preheating, from $\tau = 73$ to $\tau = 79$, the time when the GWs are being sourced most actively. The *left panels* correspond to the case $\chi_i = 3.42 \times 10^{-8} M_{Pl}$, and the *right panels* to $\chi_i = 1.0 \times 10^{-8} M_{Pl}$. The color coding is fixed during the evolution, though different between the two cases. However, the range of χ values covered by the axis is the same in both cases, such that different colours describe the same magnitude of difference in both cases. The correlation between the dynamics of the sources and the amplitude of the GWs is clearly demonstrated by this sequence of snapshots: the gradients for $\chi_i = 3.42 \times 10^{-8} M_{Pl}$ are larger than for $\chi_i = 1.0 \times 10^{-8} M_{Pl}$, in correspondence with the higher total amplitude of GWs, $\Omega_{GW} = 1.1 \times 10^{-3}$ and $\Omega_{GW} = 5.6 \times 10^{-4}$

3.5 Anisotropies in the GW Background from Massless Preheating

I will now present the final result of this chapter: The variation of GW energy from preheating on cosmological scales. To do this, it will first be necessary to introduce the mathematical machinery used to analyse the large scale anisotropy in Sect. 3.5.1.

In 3.5.2, I will quantify the relative anisotropy and show that it is of the order of 1 %. I will finish by describing how the field dynamics affect the gravitational wave production during massless preheating in Sect. 3.5.3.

3.5.1 Toolkit for Computing Anisotropies

The amount of GW production strongly depends on the value of χ_i, as I have shown explicitly in Sect. 3.4 for two values of χ_i. In the next subsection, I will present the data from many simulations, each with a different χ_i amplitude drawn from the appropriate random distribution. The dependence of Ω_{GW} on χ_i turns out to be very irregular, see Fig. 3.10. We will need to perform a statistical analysis of the data to extract the anisotropy from the $\Omega_{\mathrm{GW}}(\chi_\mathrm{i})$ dependence obtained from the simulations. Hence, in the following I will provide a mathematical toolkit for such an analysis.

To begin with, let us assume a situation where $\Omega_{\mathrm{GW}}(\chi_\mathrm{i})$ depends linearly on χ_i. We will not need this to be the case in general (and as I mentioned, the dependence is actually very irregular), but it will be instructive to study the linear relation as a starting point. Normalizing the χ_i variations to the natural scale of the problem, H_*, we can then write

$$\Omega_{\mathrm{GW}}(\chi_\mathrm{i}) = c_0 + c_1 \frac{\delta\chi_\mathrm{i}}{H_*}, \tag{3.46}$$

with $\delta\chi_\mathrm{i} \equiv \chi_\mathrm{i} - \overline{\chi}_\mathrm{i}$, where $\overline{\chi}_\mathrm{i}$ is the mean value over the currently observable universe. The constants c_0, c_1 are dimensionless and completely characterise the function $\Omega_{\mathrm{GW}}(\chi_\mathrm{i})$ (under the linear assumption). From Eq. (3.46) one can easily see that c_0 can be identified with the mean amplitude of the GWs over the observable universe, $c_0 \equiv \overline{\Omega}_{\mathrm{GW}}$. We can then express the relative fluctuations of the GW energy density as

$$\delta\Omega_{\mathrm{GW}} \equiv \frac{\Omega_{\mathrm{GW}} - \overline{\Omega}_{\mathrm{GW}}}{\overline{\Omega}_{\mathrm{GW}}} \equiv \frac{c_1}{c_0}\frac{\delta\chi_\mathrm{i}}{H_*}. \tag{3.47}$$

As these fluctuations are proportional to $\delta\chi_\mathrm{i}$, like χ_i they represent a nearly Gaussian and scale-invariant random field. The power spectrum of $\delta\Omega_{\mathrm{GW}}$ can then be directly related to the power spectrum $\mathcal{P}_\chi$ of χ_i by

$$\mathcal{P}_{\mathrm{GW}} = \frac{c_1^2}{c_0^2}\frac{\mathcal{P}_\chi}{H_*^2} = \frac{1}{4\pi^2}\frac{c_1^2}{c_0^2}, \tag{3.48}$$

where we used Eq. (3.26). To measure fluctuations on the celestial sphere, it is better to express them in terms of spherical harmonics $\{Y_{lm}\}$. This makes it possible to characterise the statistical properties of $\delta\Omega_{\mathrm{GW}}$ in terms of an angular power spectrum, in the same way as one does for the CMB temperature anisotropies. We can decompose the fluctuations in the GW energy density as

$$\delta\Omega_{\rm GW}(\hat{n}) = \sum_{l\geq 1}^{\infty}\sum_{m=-l}^{+l} g_{lm}\mathrm{Y}_{lm}(\hat{n}), \tag{3.49}$$

where $g_{lm} = \int_{4\pi} d\Omega \mathrm{Y}^*_{lm}(\hat{n})\delta\Omega_{\rm GW}(\hat{n})$ are (complex) coefficients weighting each angular moment. The angular power spectrum C_l is defined as the ensemble average of the coefficients,

$$\left\langle g^*_{lm} g_{l'm'} \right\rangle \equiv C_l \delta_{ll'}\delta_{mm'}, \tag{3.50}$$

where the Kronecker delta $\delta_{ll'}\delta_{mm'}$ and the dependence of C_l on only l reflects statistical isotropy. The C_l's are given by

$$C_l \equiv 2\pi \int d\cos\theta P_l(\cos\theta) C(\cos\theta), \tag{3.51}$$

where $P_l(\cos\theta)$ are the Legendre polynomials, and $C(\cos\theta)$ is the angular correlation of the GW fluctuations at different directions in the sky $\hat{n}_1$ and $\hat{n}_2$:

$$C(\cos\theta) \equiv \left\langle \delta\Omega_{\rm GW}(\hat{n}_1)\delta\Omega_{\rm GW}(\hat{n}_2)\right\rangle, \tag{3.52}$$

with $\hat{n}_1 \cdot \hat{n}_2 \equiv \cos\theta$.

Equivalently, the angular correlation can be expressed as a linear sum in the C_l's weighted as

$$\left\langle \delta\Omega_{\rm GW}(\hat{n}_1)\delta\Omega_{\rm GW}(\hat{n}_2)\right\rangle = \sum_{l\geq 1}^{\infty} \frac{(2l+1)}{4\pi} C_l\, P_l(\cos\theta). \tag{3.53}$$

Because of the assumed linear relation between $\delta\Omega_{\rm GW}$ and $\delta\chi_{\rm i}$ in Eq. (3.47), the angular power spectrum of the GW energy density fluctuations can be calculated very easily.

Deriving the angular power spectrum C_l from the linear power spectrum $\mathcal{P}_{\rm GW}$ is a standard exercise which is performed in e.g. [25], where it is used to compute the temperature power spectrum on large angular scales (which corresponds to the Sachs-Wolfe plateau). The relation is simply given by

$$l(l+1)C_l = \frac{\pi}{2}\mathcal{P}_{\rm GW} = \frac{1}{8\pi}\frac{c_1^2}{c_0^2}. \tag{3.54}$$

When this calculation is performed for small l for the CMB fluctuations, it demonstrates that on very large angular scales the power spectrum looks approximately flat when multiplied by $l(l+1)$. This reflects the scale invariance of the primordial power spectrum from inflation, as the largest scales were superhorizon at the time of recombination and therefore had not evolved much. In our case, GWs decouple upon production and do not evolve inside the horizon, and the relation holds on all angular scales.

From (3.54) we can see that as long as $\delta\Omega_{\rm GW}$ is linearly dependent on $\delta\chi_{\rm i}$ as in Eq. (3.46), the coefficients c_0 and c_1 completely determine the angular power spectrum. In the case of massless preheating we are considering, and generally in any other scenario of preheating, the $\Omega_{\rm GW}(\chi_{\rm i})$ relationship will not be linear. In [33] it was discussed how to derive the angular power spectrum under these circumstances, which motivated the approach I am going to describe.

To describe fluctuations on any angular scale independent of the functional form of the relation $\Omega_{\rm GW}(\chi_{\rm i})$, we need to compute the two-point correlation function of the GW energy density originating from two points $\mathbf{x}$ and $\mathbf{y}$. Due to isotropy, this correlator can only depend on the separation $|\mathbf{x}-\mathbf{y}|$. It can be written as

$$\left\langle\Omega_{\rm GW}(\mathbf{x})\Omega_{\rm GW}(\mathbf{y})\right\rangle \equiv \int d\chi_{\mathbf{x}}d\chi_{\mathbf{y}}P(\chi_{\mathbf{x}},\chi_{\mathbf{y}})\Omega_{\rm GW}(\chi_{\mathbf{x}})\Omega_{\rm GW}(\chi_{\mathbf{y}})\,, \tag{3.55}$$

where $P(\chi_{\mathbf{x}},\chi_{\mathbf{y}})$ is the joint probability distribution for the field values $\chi_{\mathbf{x}}=\chi_{\rm i}(\mathbf{x})$ and $\chi_{\mathbf{y}}=\chi_{\rm i}(\mathbf{y})$ at the points $\mathbf{x}$ and $\mathbf{y}$. Since these are Gaussian random fields, we have

$$P(\chi_{\mathbf{x}},\chi_{\mathbf{y}}) = \frac{1}{2\pi\sqrt{|G|}}\,e^{-\frac{1}{2}\vec{\delta\chi}^{\rm T}G^{-1}\vec{\delta\chi}}\,, \tag{3.56}$$

where I defined the vector $\vec{\delta\chi} \equiv (\chi_{\mathbf{x}}-\overline{\chi}_{\rm i},\chi_{\mathbf{y}}-\overline{\chi}_{\rm i})$. The 2×2 covariant matrix G and its inverse G^{-1}, with determinant $|G|$, are given by

$$G \equiv \begin{pmatrix} G_{\mathbf{x},\mathbf{x}} & G_{\mathbf{x},\mathbf{y}} \\ G_{\mathbf{x},\mathbf{y}} & G_{\mathbf{y},\mathbf{y}} \end{pmatrix}\,,\quad G^{-1} \equiv \frac{1}{|G|}\begin{pmatrix} G_{\mathbf{y},\mathbf{y}} & -G_{\mathbf{x},\mathbf{y}} \\ -G_{\mathbf{x},\mathbf{y}} & G_{\mathbf{x},\mathbf{x}} \end{pmatrix}\,, \tag{3.57}$$

with $G_{\mathbf{x},\mathbf{y}} \equiv \langle\delta\chi_{\rm i}(\mathbf{x})\delta\chi_{\rm i}(\mathbf{y})\rangle$ the field correlator, and $\sigma_\chi^2 = G_{\mathbf{x},\mathbf{x}} = \langle\delta\chi^2\rangle$ the field variance, which is given by Eq. (3.28). From the scale-invariant power spectrum (3.26), on sufficiently large scales (ignoring the oscillating factor $\exp\left[-i\mathbf{k}\cdot(\mathbf{x}-\mathbf{y})\right]$) we can approximate the 2-point function as

$$G_{\mathbf{x},\mathbf{y}} \approx \frac{H_*^2}{4\pi^2}\Big[\ln\left(a_0/|\mathbf{x}-\mathbf{y}|\right)-\ln(a_0H_0)\Big] = \frac{H_*^2}{4\pi^2}\ln\Big[\left(|\mathbf{x}-\mathbf{y}|H_0\right)^{-1}\Big], \tag{3.58}$$

where I integrated from horizon scales a_0H_0, with H_0 the Hubble rate today, up to comoving subhorizon scales $k=a_0/|\mathbf{x}-\mathbf{y}|$.

Scales larger than the Hubble volume are not considered, as we are evaluating the 2-point function of fluctuations around the mean value $\bar{\chi}_{\rm i}$ across the observable universe. The correlator therefore goes to zero as we approach Hubble scales $|\mathbf{x}-\mathbf{y}|\sim 1/H_0$, and is only well defined down to scales $|\mathbf{x}-\mathbf{y}|\sim a_0/a_*H_*$ of the order of the Hubble horizon at the end of inflation, for which the expression in Eq. (3.58) reduces

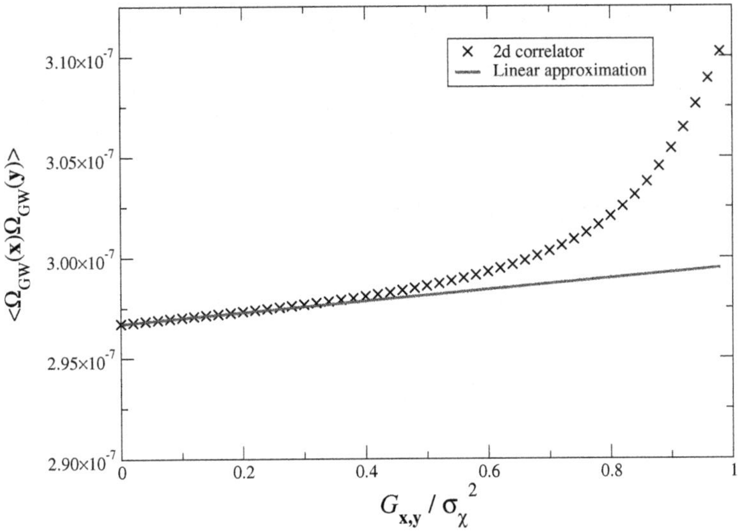

Fig. 3.7 The full GW energy density correlator and its linearised version. The two results agree very well on the largest currently observable scales, i.e. for small $G_{\mathbf{x},\mathbf{y}}/\sigma_\chi^2$ values

to the field variance σ_χ^2 [which has the same high momentum cutoff $k = H_*$, see Eq. (3.28)].

By obtaining the function $\Omega_{\rm GW}(\chi_{\rm i})$ from lattice simulations, I computed the GW energy density correlator (3.55) numerically. This is shown in Fig. 3.7 for $\chi_{\rm i} = 3.42 \times 10^{-7} M_{\rm Pl}$. Note that the correlator only depends on the distance $|\mathbf{x} - \mathbf{y}|$ through the ratio $G_{\mathbf{x},\mathbf{y}}/G_{\mathbf{x},\mathbf{x}} = G_{\mathbf{x},\mathbf{y}}/\sigma_\chi^2$.

In principle, one could use the numerical solution to compute the angular correlation of the GW energy density between any two directions $\hat{n}_1, \hat{n}_2$ in the sky by evaluating Eq. (3.55) at positions $\mathbf{x} = R\hat{n}_1$ and $\mathbf{y} = R\hat{n}_2$, with $R \sim H_0^{-1}$ the distance to the 'scattering surface' at preheating where the GWs were emitted. From there we could obtain the angular power spectrum C_l by means of Eq. (3.51).

In practice, this procedure can be cumbersome and, more importantly, since $\Omega_{\rm GW}(\chi_{\rm i})$ may be very irregular, it would be difficult to assess the accuracy in the final amplitude of the C_l's. Instead, I will make use of the fact that on large scales $|\mathbf{x} - \mathbf{y}| \lesssim 1/H_0$, for which the logarithm in Eq. (3.58) is less than unity, the ratio

$$\frac{G_{\mathbf{x},\mathbf{y}}}{\sigma_\chi^2} \lesssim \frac{1}{N_*} \simeq 0.017\,, \tag{3.59}$$

is very small. As Fig. 3.7 shows, the correlator on these scales is very well approximated by a linear function. This is because on large scales we average over the small, irregular fluctuations and only retain the smooth, underlying functional dependence.

To simplify the analysis and to avoid having to compute the full correlation function Eq. (3.55), we can therefore perform a linear Taylor expansion of the joint probability distribution in powers of the field correlator normalized to the variance, $G_{\mathbf{x},\mathbf{y}}/\sigma_\chi^2$. This gives

$$P(\chi_{\mathbf{x}}, \chi_{\mathbf{y}}) \propto \exp\left\{ -\frac{(\delta\chi_{\mathbf{x}}^2 + \delta\chi_{\mathbf{y}}^2) - 2\left(\frac{G_{\mathbf{x},\mathbf{y}}}{\sigma_\chi^2}\right)\delta\chi_{\mathbf{x}}\delta\chi_{\mathbf{y}}}{2\sigma_\chi^2\left[1 - \left(\frac{G_{\mathbf{x},\mathbf{y}}}{\sigma_\chi^2}\right)^2\right]} \right\} \tag{3.60}$$

$$\approx \exp\left(-\frac{\delta\chi_{\mathbf{x}}^2}{2\sigma_\chi^2}\right)\exp\left(-\frac{\delta\chi_{\mathbf{y}}^2}{2\sigma_\chi^2}\right)\left[1 + \left(\frac{G_{\mathbf{x},\mathbf{y}}}{\sigma_\chi^2}\right)\frac{\delta\chi_{\mathbf{x}}\delta\chi_{\mathbf{y}}}{\sigma_\chi^2} + \mathcal{O}\left(\frac{G_{\mathbf{x},\mathbf{y}}}{\sigma_\chi^2}\right)^2\right].$$

Substituting this expansion into Eq. (3.55), we then obtain

$$\langle\Omega_{\rm GW}(\mathbf{x})\Omega_{\rm GW}(\mathbf{y})\rangle \simeq \langle\Omega_{\rm GW}(\chi_{\rm i})\rangle^2 + \frac{\langle\delta\chi_{\rm i}\Omega_{\rm GW}(\chi_{\rm i})\rangle^2}{\sigma_\chi^2}\left(\frac{G_{\mathbf{x},\mathbf{y}}}{\sigma_\chi^2}\right) + \mathcal{O}\left(\frac{G_{\mathbf{x},\mathbf{y}}}{\sigma_\chi^2}\right)^2, \tag{3.61}$$

where the expectation values on the right hand side are given by

$$\langle\Omega_{\rm GW}\rangle \equiv \int d\chi_{\rm i} P(\chi_{\rm i})\Omega_{\rm GW}(\chi_{\rm i}), \tag{3.62}$$

$$\langle\delta\chi_{\rm i}\Omega_{\rm GW}(\chi_{\rm i})\rangle \equiv \int d\chi_{\rm i} P(\chi_{\rm i})\delta\chi_{\rm i}\Omega_{\rm GW}(\chi_{\rm i}), \tag{3.63}$$

which need to be computed using the single-point probability distribution

$$P(\chi_{\rm i}) = \frac{1}{\sqrt{2\pi}\,\sigma_\chi}\exp\left\{-\frac{1}{2}\frac{(\chi_{\rm i} - \overline{\chi}_{\rm i})^2}{\sigma_\chi^2}\right\}. \tag{3.64}$$

Re-arranging the terms on the right-hand side of Eq. (3.61), we can write the equation as

$$\langle\Omega_{\rm GW}(\mathbf{x})\Omega_{\rm GW}(\mathbf{y})\rangle \simeq \left\langle\left(\langle\Omega_{\rm GW}\rangle + \frac{\langle\delta\chi_{\rm i}\Omega_{\rm GW}(\chi_{\rm i})\rangle}{\sigma_\chi^2}\delta\chi_{\rm i}(\mathbf{x})\right)\left(\langle\Omega_{\rm GW}\rangle + \frac{\langle\delta\chi_{\rm i}\Omega_{\rm GW}(\chi_{\rm i})\rangle}{\sigma_\chi^2}\delta\chi_{\rm i}(\mathbf{y})\right)\right\rangle. \tag{3.65}$$

This is precisely the form of Eq. (3.46), which was derived for a linear relationship between the fluctuations in GW amplitude and those in the field χ. On large scales, the linear Taylor expansion is a very good approximation to the full function, and hence the expression obtained for the angular power spectrum derived for the linear relation,

Eq. (3.54), should be valid for any functional dependence $\Omega_{\rm GW}(\chi)$. Identifying the coefficients as

$$c_0 = \langle \Omega_{\rm GW} \rangle\,, \qquad c_1 = \frac{H_*}{\sigma_\chi^2} \left\langle \delta\chi_{\rm i} \Omega_{\rm GW}(\chi_{\rm i}) \right\rangle\,, \tag{3.66}$$

we can use Eq. (3.54) directly to compute the angular power spectrum of the relative GW energy density fluctuations:

$$l(l+1)C_l = \frac{H_*^2}{8\pi} \frac{\langle \delta\chi_{\rm i} \Omega_{\rm GW}(\chi_{\rm i})\rangle^2}{\sigma_\chi^4 \langle \Omega_{\rm GW}\rangle^2}\,. \tag{3.67}$$

This equation is one of the main results of this chapter. It is a master formula for the angular power spectrum of the energy density fluctuations of any GW background of cosmological origin, whose anisotropies originated from the modulation due to an inflationary spectator field. It is valid on large angular scales for which the Taylor expansion holds, which in any case dominate over small scales as the spectrum decays as $C_l \sim 1/l^2$. If detectable, the effect would therefore probably be easiest to measure on the level of the quadrupole ($l = 2$), as the dipole might be dominated by the motion of our galaxy.

We are now able to calculate the typical amplitude of fluctuations $\sim \sqrt{l(l+1)C_l}$ for any value of $\overline{\chi}_{\rm i}$, by simply evaluating the expectation values in Eq. (3.67) from the results of a lattice simulation numerically.

3.5.2 Anisotropic Gravitational Wave Background

According to the considerations in Sect. 3.2.2, I chose a mean value $\overline{\chi}_{\rm i} = 3.42 \times 10^{-7} M_{\rm Pl}$ across our observable universe, and a variance $\sigma_\chi^2 = 3.3 \times 10^{-15} M_{\rm Pl}^2$ within our current Hubble volume, to describe the range of initial $\chi_{\rm i}$ values the GW background from preheating is likely to have originated from.

Following the Monte Carlo method [34], I randomly chose $\mathcal{N} = 500$ initial values $\chi_{\rm i}^j$, $j \in \{1, \ldots, \mathcal{N}\}$ from the Gaussian distribution (3.64). To be exact, I randomly picked 250 values $\chi_{\rm i}^j$ and chose the remaining half to be the symmetric value $\chi_{\rm i}^{j'} = 2\overline{\chi}_{\rm i} - \chi_{\rm i}^j$. This ensures that the mean of the distribution will be exactly the required $\overline{\chi}_{\rm i} = 3.42 \times 10^{-7} M_{\rm Pl}$, which reduces the error when computing expectation values. This is necessary as there is only a finite sample of values so we would never be able to obtain a perfect Gaussian distribution.

The Monte Carlo method has the advantage of making it easier to sample the highly chaotic variation of the GW energy density $\Omega_{\rm GW}(\chi_{\rm i}^j)$ without needing to use a very small step size in $\chi_{\rm i}$, as well as simplifying the computation of the expectation values in Eq. (3.67).

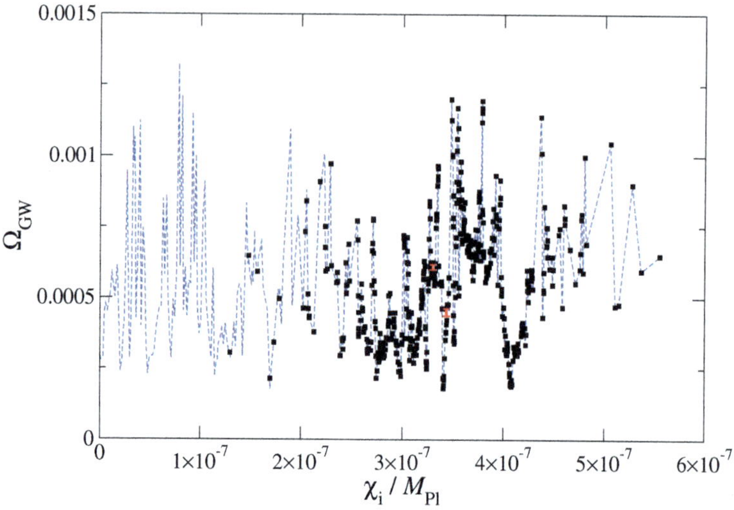

Fig. 3.8 Ω_{GW} for our sample of initial field values χ_i. The squares show the Gaussian ensemble used for the analysis. The *red error bars* show the standard deviation between different seed values for the same χ_i. The *dotted line* shows the GW energy density for a wider range of χ_i for illustration purposes

For each χ_i^j, I performed one simulation run and evaluated the GW energy density $\Omega_{GW}(\chi_i^j)$, see Fig. 3.8. The black squares show the 500 initial values picked during the Monte Carlo simulation, while the blue dotted line includes some smaller values of χ_i to show the dependence over a larger range.

As the plot illustrates, Ω_{GW} is highly dependent on χ_i, varying by as much as a factor of five between nearby values, although there are some ranges of χ_i where the dependence is much smoother. This irregular behaviour is in line with the chaotic field dynamics observed when studying curvature perturbations [27, 28, 35], but the amplitude of fluctuations is unexpectedly high.

The figure also shows the variation of Ω_{GW} due to different random realisations of the field fluctuations for two initial χ_i, illustrated by the tiny red error bars. The magnitude of the error in these two cases is representative of the statistical error in the GW energy density for any value of χ_i. It is clearly much smaller than the variation of Ω_{GW} between different values of χ_i, confirming that the effect is not merely statistical fluctuation.

As I used a Monte Carlo method to choose the range of χ_i^j, the expectation values in Eq. (3.67) can simply be approximated by averages over the sample,

$$\langle \Omega_{GW} \rangle \approx \frac{1}{\mathcal{N}} \sum_j \Omega_{GW}(\chi_i^j),$$

$$\langle \delta\chi\Omega_{\rm GW}\rangle \approx \frac{1}{\mathcal{N}}\sum_j (\chi_{\rm i}^j - \overline{\chi}_{\rm i})\Omega_{\rm GW}(\chi_{\rm i}^j). \tag{3.68}$$

Averaging over the 500 data points in Fig. 3.8 which correspond to $\overline{\chi}_{\rm i} = 3.42 \times 10^{-7} M_{\rm Pl}$, I obtained $\langle \Omega_{\rm GW}\rangle = (5.45 \pm 0.13) \times 10^{-4}$ and $\langle \delta\chi\Omega_{\rm GW}\rangle = (3.0 \pm 1.2) \times 10^{-12} M_{\rm Pl}$. Substituting these into Eq. (3.67) gives the amplitude of the angular power spectrum of the relative fluctuations $\delta\Omega_{\rm GW} = (\Omega_{\rm GW}/\overline{\Omega}_{\rm GW} - 1)$ as

$$\sqrt{l(l+1)C_l} = 0.017 \pm 0.003, \tag{3.69}$$

where the errors are estimated by the bootstrap method [36]. This method provides a useful way of measuring the uncertainty in expectation values calculated from a single data set, by mimicking the process of obtaining new data from the same probability distribution.

Assuming there are N data points in the original ensemble, for each bootstrap sample N of these points are randomly selected, without avoiding double counting. The expectation value is then calculated based on the current set of data points, and the variance of many such bootstrap samples gives an estimate of the error in the expectation value.

In this case, I used 1000 bootstrap samples of 250 randomly chosen symmetric pairs $\chi_{\rm i}^{j'}, \chi_{\rm i}^{j}$ (to make sure that each bootstrap sample has the correct mean $\overline{\chi}_{\rm i} = 3.42 \times 10^{-7} M_{\rm Pl}$) to calculate (3.68), and the variance of these samples gave an error estimate of magnitude 0.003 for the amplitude of relative fluctuations.

The set of initial $\chi_{\rm i}$ was generated by the Python random number generator, using a Gaussian probability distribution with the chosen mean and variance. For unknown reasons, the numerical value of the variance of the data set turned out to be $\sigma^2 = 4.3 \times 10^{-15} M_{\rm Pl}^2$, which is significantly higher than the desired value $\sigma_\chi^2 = 3.3 \times 10^{-15} M_{\rm Pl}^2$. To rectify this, I reweighted the data to resemble a sample with a variance closer to the required one.

Reweighting [37] makes it possible to use Monte Carlo data belonging to a specific probability distribution to calculate expectation values for other, similar distributions. Assume values x were drawn from a probability distribution $p(x)$ and you need to calculate the expectation value of an observable O from a slightly different probability distribution function $p'(x)$,

$$\langle O\rangle' = \frac{\int dx p'(x) O(x)}{\int dx p'(x)}. \tag{3.70}$$

We can re-express this in terms of the old probability distribution $p(x)$ as

$$\langle O\rangle' = \frac{\int dx p(x)\frac{p'(x)}{p(x)} O(x)}{\int dx p(x)\frac{p'(x)}{p(x)}} = \frac{\langle r(x) O(x)\rangle}{\langle r(x)\rangle} \equiv \frac{\sum_j r(x_j) O(x_j)}{\sum_j r(x_j)}, \tag{3.71}$$

where in the last step I used the fact that expectation values are simply sums for a Monte Carlo data set, and where $r(x) = \frac{p'(x)}{p(x)}$ is the reweighting factor. Therefore, to calculate expectation values from a slightly different probability distribution to the original one, we can simply reweight each observable by $r(x_j)$. As long as the probability distributions are close to each other, i.e. $\frac{1}{N}\sum_j r(x_j) \approx 1$, this method can be trusted.

The numerical data presented in Fig. 3.8 suggests a Gaussian probability distribution with mean $\overline{\chi}_\mathrm{i}$ and variance σ^2. This needs to be reweighted to obtain the correct variance $\sigma_\chi^2 = 3.3 \times 10^{-15} M_\mathrm{Pl}^2$, so we have to evaluate the expectation value (3.71) for the probability distribution

$$p'(\chi_\mathrm{i}) = \frac{1}{\sqrt{2\pi\sigma_\chi^2}} \exp\left(-\frac{(\chi_\mathrm{i} - \overline{\chi}_\mathrm{i})^2}{2\sigma_\chi^2}\right). \tag{3.72}$$

Note that the reweighting takes place for the whole sample (when calculating the mean expectation value) and for each bootstrap sample (when estimating the errors).

By employing the method of reweighting, we can also use the Monte Carlo data to calculate expectation values around different nearby mean values $\overline{\chi}'_\mathrm{i}$ (which corresponds to a χ background with a slightly different average across our observable universe). These values will have a Gaussian probability distribution

$$p'(\chi_\mathrm{i}) = \frac{1}{\sqrt{2\pi\sigma_\chi^2}} \exp\left(-\frac{(\chi_\mathrm{i} - \overline{\chi}'_\mathrm{i})^2}{2\sigma_\chi^2}\right), \tag{3.73}$$

where $\overline{\chi}'_\mathrm{i}$ is a different mean value to the one chosen in the simulations. The total reweight factor is therefore

$$r(\chi_\mathrm{i}^j) = \frac{\sigma}{\sigma_\chi} \exp\left[-\frac{(\chi_\mathrm{i}^j - \overline{\chi}'_\mathrm{i})^2}{2\sigma_\chi^2} + \frac{(\chi_\mathrm{i}^j - \overline{\chi}_\mathrm{i})^2}{2\sigma^2}\right]. \tag{3.74}$$

We can use this procedure to calculate C_l from the expectation values in (3.68), evaluated around the new probability distribution by use of Eq. (3.71). The solid line in Fig. 3.9 shows the relative amplitude of angular fluctuations for different mean values $\overline{\chi}_\mathrm{i}$ across our observable universe, where the red dot corresponds to our original choice $\overline{\chi}_\mathrm{i} = 3.42 \times 10^{-7} M_\mathrm{Pl}$. For the reweighted mean values, the error bars in the fluctuations have been obtained by the bootstrap method, similarly to the original value.

One point to note about the plot is that the reweighted data sets have an uncertainty in the value of $\overline{\chi}'_\mathrm{i}$, because once reweighted, the actual mean of each bootstrap sample (all of which have a mean $\overline{\chi}_\mathrm{i} = 3.42 \times 10^{-7} M_\mathrm{Pl}$ before reweighting, due to the choice of symmetric pairs) is slightly different to $\overline{\chi}'_\mathrm{i}$, as we only have a finite number of data points. The uncertainty in $\overline{\chi}'_\mathrm{i}$ becomes larger far away from the original mean,

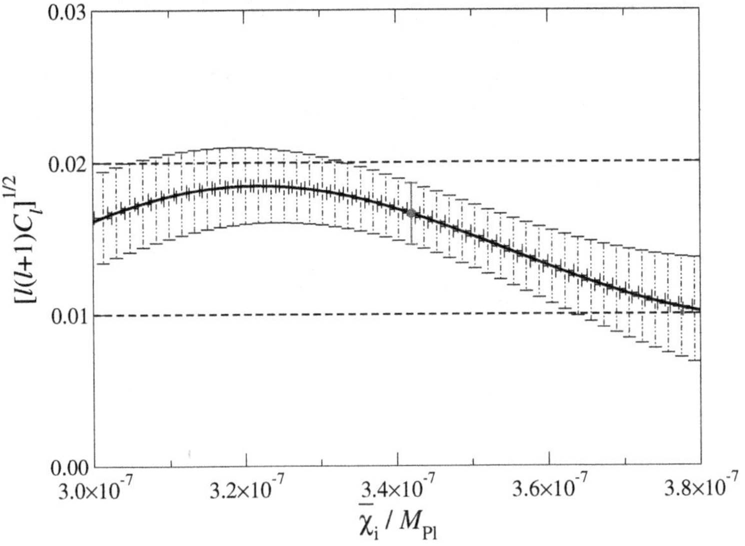

Fig. 3.9 The relative amplitude of the multipoles of the GW background as a function of the average field value $\overline{\chi}_i$, calculated from Eq. (3.67). The *red dot* shows the amplitude for original mean value $\overline{\chi}_i = 3.42 \times 10^{-7} M_{Pl}$, and the *curve* shows values obtained by reweighting the same data

where we do not have enough coverage to simulate a probability distribution with the chosen new mean. This is shown by the horizontal error bars in Fig. 3.9.

Due to this uncertainty, when computing expectation values as in Eq. (3.68), the value of the chosen mean $\overline{\chi}_i'$ was actually replaced by the reweighted value, which ensures that the Monte Carlo averages are performed over the actual mean of the sample. Again, this was done for the whole sample and each bootstrap sample separately. Although I am taking this subtlety into account when calculating the angular power spectrum, ignoring it (and simply using the chosen means $\overline{\chi}_i'$ in all calculations) does not make a substantial difference to the final result, as the shift in mean value due to reweighting is small.

For most of the range of $\overline{\chi}_i$ presented in Fig. 3.9, the amplitude of the fluctuations is above the one percent level, even within error bars. This is much higher than the relative amplitude of fluctuations in the CMB which is of order 10^{-5}. If the fluctuations in the GW background had been tiny, we would never be able to detect them. However, it is reasonable to hope that variations of order 1 % could be measured by future GW detectors, although it is very hard to make any statement about their sensitivity to anisotropies at the current stage.

Unfortunately, as mentioned at the beginning of this chapter, even before considering anisotropies, GWs from preheating are not within the sensitivity range of the current main detectors, see Fig. 3.1. Let me demonstrate this for the results obtained from my simulations. Using Eq. (3.22) and the typical dimensionless peak GW frequency $\kappa = 0.2$ as shown in Fig. 3.5, we can obtain the frequency today:

$$f \approx 0.2 \times (9 \cdot 10^{-14})^{1/4} \times 7 \cdot 10^{10}\text{Hz} \approx 7.7\,\text{MHz}\,. \tag{3.75}$$

From Fig. 3.8, we can see that the average energy density is approximately $\Omega_{\text{GW}} = 0.00075$, and hence its value today is

$$h^2 \Omega^0_{\text{GW}} \approx 6.8 \cdot 10^{-9}\,. \tag{3.76}$$

This energy density corresponds to the upper end of values in Fig. 3.1. This shows that the value $g^2/\lambda = 2$ leads to the production of a very large number of GWs, which likely is related to the amplification of long wavelength modes which transfer a lot of power into the field.

3.5.3 Field Dynamics

As described in Sect. 3.2.2, the mean value of χ across our observable universe, $\overline{\chi}_{\text{i}}$, is a free parameter dependent on the total number of e-folds of inflation. To have a complete picture of the anisotropies in the GW background, one should therefore analyse a wider range of χ_{i} values than the one considered in the Monte Carlo simulation.

In Fig. 3.8, I have already included the GW energy density for smaller values of χ_{i}, and this data is reproduced in Fig. 3.10 on a logarithmic scale for illustrative purposes. As it is reasonable to assume inflation lasted some number of e-folds longer than the minimal required number of $N_* = 60$, much smaller values of χ_{i} than presented in Fig. 3.10 are very unlikely, as even with $N_{\text{tot}} = 70$ we would have at least an expected value of order $\overline{\chi}_{\text{i}} \sim 3 \times 10^{-8} M_{\text{Pl}}$, see Eq. (3.30).

On the other hand, there could definitely be larger values of χ_{i} in our observable universe if inflation lasted for a very long time. Due to limitations in computing power (each run took about six hours on 64 processors, and I had to perform several hundreds of them), I chose not to perform any simulations for larger values of χ_{i}.

The data presented in Fig. 3.10 reveals some non-trivial structure. In particular, the GW energy density has an approximate log-periodic dependence on χ_{i}, with regions of high, quickly varying GW amplitude alternating with regions of low amplitude. To make this more apparent, I have also included a curve that shows the convolution of the data with a Gaussian window function,

$$\widetilde{\Omega}_{\text{GW}}(\log\chi) = \frac{1}{\sqrt{2\pi\sigma^2_{\text{w}}}} \int d\delta\, e^{-\delta^2/2\sigma^2_{\text{w}}}\, \Omega_{\text{GW}}(\log\chi + \delta)\,, \tag{3.77}$$

where $\sigma^2_{\text{w}} = 0.05$ is the spread of the window function. A log-periodic structure in the field dynamics was predicted by [35] in the context of studying curvature perturbations from massless preheating, and I will comment on it again at the end of this section.

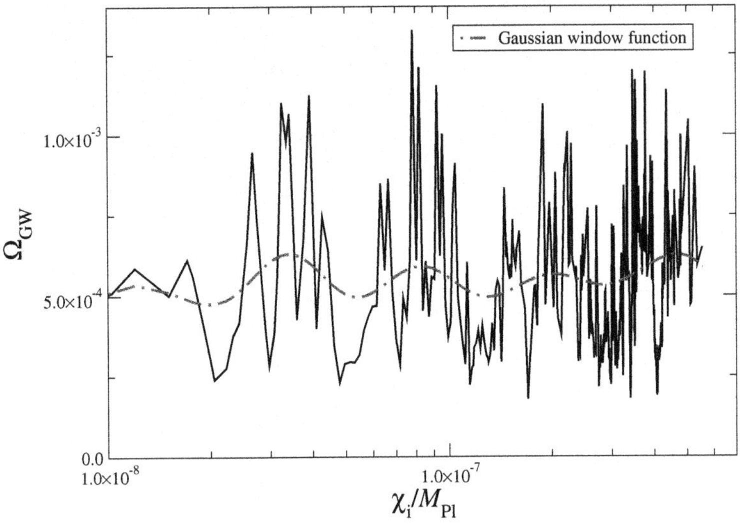

Fig. 3.10 $\Omega_{\rm GW}$ plotted with logarithmic $\chi_{\rm i}$ axis. The *blue dot-dashed curve* corresponds to the convolution of the data with a Gaussian window function to make the periodic structure in $\log\chi_{\rm i}$ more apparent

First, let me further elucidate the physical origin of the sensitive dependence of the GW amplitude on $\chi_{\rm i}$, by studying the relationship between GW production and field dynamics. As the source term for tensor perturbations is given by the field gradients, four powers of which will appear in the equation for the GW power spectrum (3.18), it is natural to ask which of the scalar fields is primarily responsible for the production of GWs.

In Fig. 3.11 I have plotted the total power spectrum of GWs, as well as the power spectrum obtained from using only ϕ or χ as a source of GWs (the total amplitude will also contain cross terms between the fields). The plot shows that the GWs are sourced primarily by the gradients of the χ field, which is not surprising, as the inflaton fluctuations do not get amplified very strongly as we saw in Fig. 3.2. We can therefore focus on the dynamics of χ in order to understand the physical origin of the variation of the GW energy density.

In Ref. [35] it was observed that the evolution of the system during massless preheating strongly depends on the relative phase of the homogeneous modes $\phi(t)$ and $\chi(t)$ at the time the field dynamics become non-linear (i.e. when χ becomes sufficiently large).

In particular, in some cases $\chi(t)$ acquires a very large amplitude compared to the inflaton, leading to a spiky contribution to the curvature. Assume the inflaton oscillates with period T during the linear stage. Initial χ configurations related by

$$\frac{\chi_{\rm i}'}{\chi_{\rm i}} = e^{\mu n T}, \tag{3.78}$$

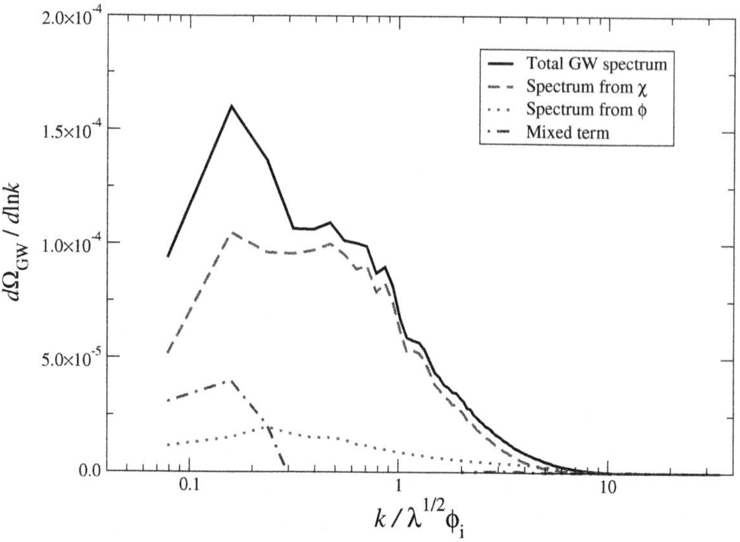

Fig. 3.11 The GW amplitude from two different χ_i sourced by only ϕ, χ and both fields respectively

where n is an integer, will then evolve similarly, as the inflaton will have the same phase at the time the system becomes non-linear (remember $\chi(t) \propto e^{\mu t}\chi_i$). In fact, if there were no inhomogeneous modes at all, the behaviour of the fields would be exactly the same for all χ_i', χ_i related as in Eq. (3.78), as in this case only the phase information matters.

As at the onset of non-linearities the inhomogeneous modes are still small, we expect the field behaviour (and therefore the value of physical observables that depend on it) to repeat periodically in the space of initial values χ_i. This was indeed observed for curvature perturbations in [35]. I have found the same effect, but in the GW amplitude: Regions of high GW amplitude repeat log-periodically, as shown in Fig. 3.10.

To quantify how the GW production and the dynamics of χ are related, I studied how the maximum value the homogeneous field χ reaches during its evolution, $\chi_{\max}$, correlates with the amplitude of the final GW background. Indeed, $\chi_{\max}$ varies considerably between different χ_i, indicating that the field dynamics proceed differently depending on the initial value.

Obviously the GWs are not sourced by the homogeneous field itself, but rather by its inhomogeneous modes. However, the latter are directly linked to the zero mode due to the transfer of energy between them during the non-linear stage, and therefore the correlation between $\chi_{\max}$ and $\Omega_{\rm GW}$ is meaningful. In Fig. 3.12, $\chi_{\max}$ is plotted against the total amount of GW energy, for the same simulations as in Fig. 3.10.

For small $\chi_{\max} < 1 M_{\rm Pl}$, we can see a clear correlation between the field dynamics and GW production: the more energy is deposited into the χ field, the more GWs are being produced. This agrees with the findings from Fig. 3.11, showing that χ is responsible for the shape and amplitude of the GW spectra.

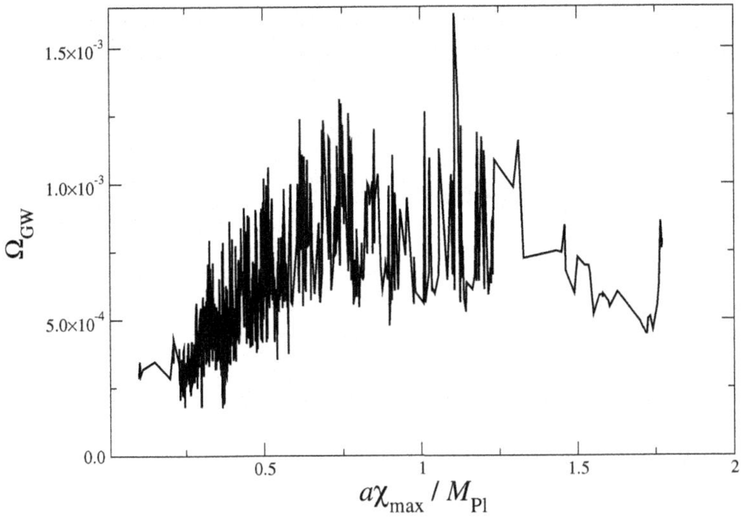

Fig. 3.12 The correlation between the maximum amplitude of the homogeneous part of χ, χ_{max}, and the total GW energy in the simulation

For high $\chi_{\text{max}} \gtrsim 1.2 M_{\text{Pl}}$, the correlation seems to turn around, and less GW are being produced, although due to the lack of data in this high χ_{max} region, it is difficult to make a proper quantitative statement. Using a smaller lattice, $\tilde{L} = 25$, I was able to find values of χ_{i} which led to a very high field value $\chi_{\text{max}} \gtrsim 5 M_{\text{Pl}}$, and for these the GW amplitude was highly suppressed.

A potential reason for the suppression might be that for low enough χ_{max}, the homogeneous $\chi(t)$ field oscillates fast enough to transfer energy to the inhomogeneous modes during the time of GW production, thus sourcing more GWs when more energy can be deposited. For very large χ_{max}, however, $\chi(t)$ only does very few oscillations, and most of the energy is stored in the homogeneous mode, thus reducing the field gradients and correspondingly the amount of GW production.

The initial values that lead to a very large amplitude of χ_{max} correspond exactly to the field behaviour that correlated with non-Gaussian curvature spikes in [35]. This suggests that a spike in the curvature is anti-correlated with the energy of gravitational waves.

However, I was only able to obtain such high values of χ_{max} for very small lattices $\tilde{L} = 25$, which do not accurately capture the peak of the GW spectrum. The reason these scenarios occur very rarely for larger lattices is that the dynamics leading to spikes are extremely sensitive to tiny variations in initial value, which are amplified if we include longer wavelength modes. To be able to study these particular values, a more accurate numerical method than the one used here is needed.

Therefore, no clear statement on the correlation between GW anisotropies and non-Gaussian features in the CMB can be made at this point. However, note that if spikes and GWs are actually anti-correlated, this effect would not be observable: The spikes only appear on very small angular scales, and therefore a suppression of the GW amplitude across such a region would be completely washed out by the large scale variation I described in the last section.

3.6 Conclusions and Outlook

Gravitational waves from preheating could provide an important tool to constrain the coupling of the inflaton to other fields and its potential in the future. I have shown that for massless preheating with a light scalar field, you would obtain an anisotropic background of GWs with relative fluctuations of the order of 1 %.

The anisotropy is a result of two separate effects: The lightness of χ and the amplification of long wavelength modes. The fact that χ is a light field is crucial for the development of anisotropy. Only if long wavelength modes have been amplified by inflation, will the initial value vary between different preheating volumes and therefore modulate the dynamics in different parts of the sky.

The second criterion ensures that the homogeneous, $k = 0$ mode is amplified by parametric resonance, and therefore the initial value χ_i will have a strong impact on the non-linear dynamics and the production of GWs.

I checked that for coupling constants g^2/λ for which $k = 0$ is not amplified strongly, e.g. $g^2/\lambda = 3, 6$, no effect was observed. This is because in these models, the high momentum fluctuations end up dominating over the homogeneous field evolution. However, this does not mean that we can only hope to observe an effect in very few, fine-tuned cases: In more general models of preheating, which contain a mass term and therefore a relevant length scale, the unstable momentum bands change with time, and typically the $k = 0$ mode is amplified for at least part of the resonance.

The strength of the anisotropy we observed clearly depends on the couplings of the model. Quantifying the anisotropy for different preheating scenarios with a light scalar field would give us new constraints on inflationary models, if we are able to observe the GW background. Primordial gravitational waves could therefore act as important probes of the early universe, alongside CMB measurements.

Although direct detection might still be a long way off, the study of CMB polarization demonstrates the constraining power of tensor fluctuations: If the result $r \approx 0.2$ from BICEP2 [22] is confirmed, it would indicate that inflation indeed happened at a very high energy (GUT) scale, and therefore simple, chaotic inflationary models with monomial potentials become more viable again. It is therefore very important to study the preheating dynamics in these scenarios.

An obvious extension to the work presented in this chapter is to consider the Higgs field as the light scalar and couple it directly to the inflaton. The existence of the Higgs has been confirmed by the LHC last year [38], with a mass $m_H \approx 126$GeV. The Higgs is a complex SU(2) doublet, $H = (h^+, h^0)$, with a potential term

$$V(H) = \lambda \left(H^\dagger H - \frac{v^2}{2} \right)^2 , \tag{3.79}$$

where $v = 246\,\text{GeV}$ the minimum, giving $\lambda = 0.13$ at tree level.

The Higgs potential depends sensitively on the running of the Higgs self coupling $\lambda(\mu)$ with energy scale μ. The measured value of the Higgs mass suggests that at at high energies the potential turns around, and λ becomes negative as we approach the Planck scale [39, 40], implying that the electroweak vacuum is metastable and there is actually a deeper vacuum at higher field values. The fact that the Higgs potential does not blow up also means that the Higgs will be a light field even at high energy scales.

Stability of the EW vacuum (which requires $\lambda > 0$ all the way up to the Planck scale) is still a possibility, however, if the value of the top mass (which has the strongest influence on the running of the self-coupling) is a few sigma away from its central value [40].

This model is therefore particularly interesting to consider, as we can take the coupling λ to be a small, positive free parameter and investigate how it affects the GW production. At high energies much larger than v, the Higgs potential (3.79) reduces to a quartic self-interaction term. Assuming a quartic inflaton potential and a quadratic coupling to the Higgs just as for the scalar field χ, this amounts to studying a very similar situation as before, but with a different, characteristic anisotropic background depending on the parameters.

This simple model ignores the coupling of the Higgs to standard model particles, whose impact on GW production was studied in [41]. For a full picture of preheating with the Higgs field, all of the couplings should be included and their impact on the GW anisotropy quantified.

References

1. J.R. Bond, A.V. Frolov, Z. Huang, L. Kofman, Non-gaussian spikes from chaotic billiards in inflation preheating. Phys. Rev. Lett. **103**, 071301 (2009). http://xxx.lanl.gov/abs/0903.3407
2. S. Hannestad, What is the lowest possible reheating temperature? Phys. Rev. **D70**, 043506 (2004). http://xxx.lanl.gov/abs/astro-ph/0403291
3. M. Maggiore, Gravitational wave experiments and early universe cosmology. Phys. Rept. **331**, 283–367 (2000). http://xxx.lanl.gov/abs/gr-qc/9909001
4. R. Easther, E.A. Lim, Stochastic gravitational wave production after inflation. JCAP **0604**, 010 (2006). http://xxx.lanl.gov/abs/astro-ph/0601617
5. L. Bethke, D.G. Figueroa, A. Rajantie, Anisotropies in the gravitational wave background from preheating. Phys. Rev. Lett. **111**(1), 011301 (2013). http://xxx.lanl.gov/abs/1304.2657
6. L. Bethke, D.G. Figueroa, A. Rajantie, On the anisotropy of the gravitational wave background from massless preheating. JCAP **1406**, 047 (2014). http://xxx.lanl.gov/abs/1309.1148
7. G.N. Felder, I. Tkachev, LATTICEEASY: A program for lattice simulations of scalar fields in an expanding universe. Comput. Phys. Commun. **178**, 929–932 (2008). http://xxx.lanl.gov/abs/hep-ph/0011159
8. L. Kofman, A.D. Linde, A.A. Starobinsky, Towards the theory of reheating after inflation. Phys. Rev. **D56**, 3258–3295 (1997). http://xxx.lanl.gov/abs/hep-ph/9704452

9. P.B. Greene, L. Kofman, A.D. Linde, A.A. Starobinsky, Structure of resonance in preheating after inflation. Phys. Rev. **D56**, 6175–6192 (1997). http://xxx.lanl.gov/abs/hep-ph/9705347
10. S.Y. Khlebnikov, I. Tkachev, Classical decay of inflaton. Phys. Rev. Lett. **77**, 219–222 (1996). http://xxx.lanl.gov/abs/hep-ph/9603378
11. T. Prokopec, T.G. Roos, Lattice study of classical inflaton decay. Phys. Rev. **D55**, 3768–3775 (1997). http://xxx.lanl.gov/abs/hep-ph/9610400
12. J. Garcia-Bellido, D.G. Figueroa, A.Sastre, A gravitational wave background from reheating after hybrid inflation. Phys. Rev. **D77**, 043517 (2008). http://xxx.lanl.gov/abs/0707.0839
13. S. Khlebnikov, I. Tkachev, Relic gravitational waves produced after preheating. Phys. Rev. **D56**, 653–660 (1997). http://xxx.lanl.gov/abs/hep-ph/9701423
14. J.F. Dufaux, A. Bergman, G.N. Felder, L. Kofman, J.-P. Uzan, Theory and numerics of gravitational waves from preheating after inflation. Phys. Rev. **D76**, 123517 (2007). http://xxx.lanl.gov/abs/0707.0875
15. K. Enqvist, D.G. Figueroa, T. Meriniemi, Stochastic background of gravitational waves from fermions. Phys. Rev. **D86**, 061301 (2012). http://xxx.lanl.gov/abs/1203.4943
16. D.G. Figueroa, T. Meriniemi, "Stochastic background of gravitational waves from fermions—theory and applications. JHEP **10**, 101 (2013). http://xxx.lanl.gov/abs/1306.6911
17. A. Cruise, R. Ingley, A prototype gravitational wave detector for 100-MHz. Class. Quant. Grav. **23**, 6185–6193 (2006)
18. A. Cruise, The potential for very high-frequency gravitational wave detection. Class. Quant. Grav. **29**, 095003 (2012)
19. T. Akutsu, S. Kawamura, A. Nishizawa, K. Arai, K. Yamamoto, et al., Search for a stochastic background of 100-MHz gravitational waves with laser interferometers. Phys. Rev. Lett. **101**, 101101 (2008). http://xxx.lanl.gov/abs/0803.4094
20. Planck Collaboration Collaboration, P. Ade et al., Planck 2013 results. XXII. Constraints on inflation. http://xxx.lanl.gov/abs/1303.5082
21. S. Tsujikawa, J. Ohashi, S. Kuroyanagi, A. De Felice, Planck constraints on single-field inflation. Phys. Rev. **D88**(2), 023529 (2013) http://xxx.lanl.gov/abs/1305.3044
22. BICEP2 Collaboration Collaboration, P.A.R. Ade et al. BICEP2 I: Detection Of B-mode Polarization at Degree Angular Scales. http://xxx.lanl.gov/abs/1403.3985
23. S. Weinberg, *Cosmology* (Oxford University Press, Oxford, 2008)
24. D.G. Figueroa, Phenomenological and Theoretical Aspects of Reheating. Ph.D. thesis
25. A.R. Liddle, D.H. Lyth, *Cosmological Inflation and Large-Scale Structure* (Cambridge University Press, Cambridge, 2000)
26. D. Wands, K.A. Malik, D.H. Lyth, A.R. Liddle, A new approach to the evolution of cosmological perturbations on large scales. Phys. Rev. **D62**, 043527 (2000). http://xxx.lanl.gov/abs/astro-ph/0003278
27. T. Tanaka, B. Bassett, Application of the separate universe approach to preheating. http://xxx.lanl.gov/abs/astro-ph/0302544
28. T. Suyama, S. Yokoyama, Generating the primordial curvature perturbations in preheating. Class. Quant. Grav. **24**, 1615–1626 (2007). http://xxx.lanl.gov/abs/astro-ph/0606228
29. A. Chambers, A. Rajantie, Lattice calculation of non-Gaussianity from preheating. Phys. Rev. Lett. **100**, 041302 (2008). http://xxx.lanl.gov/abs/0710.4133
30. A. Chambers, A. Rajantie, Non-Gaussianity from massless preheating. JCAP **0808**, 002 (2008). http://xxx.lanl.gov/abs/0805.4795
31. D.G. Figueroa, J. Garcia-Bellido, A. Rajantie, On the transverse-traceless projection in lattice simulations of gravitational wave production. JCAP **1111**, 015 (2011). http://xxx.lanl.gov/abs/1110.0337
32. WMAP Collaboration Collaboration, D. Spergel et al. First year Wilkinson Microwave anisotropy probe (WMAP) observations: determination of cosmological parameters. Astrophys. J. Suppl. **148**, 175–194 (2003). http://xxx.lanl.gov/abs/astro-ph/0302209
33. T. Suyama, S. Yokoyama, Statistics of general functions of a Gaussian field-application to non-Gaussianity from preheating. JCAP **1306**, 018 (2013). http://xxx.lanl.gov/abs/1303.1254

34. G.S. Fishman, *Monte Carlo: Concepts, Algorithms, and Applications* (Springer, New York, 1995)
35. J. Bond, G. Efstathiou, Cosmic background radiation anisotropies in universes dominated by nonbaryonic dark matter. Astrophys. J. **285**, L45–L48 (1984)
36. B. Efron, R. Tibshirani, *An Introduction to the Bootstrap* (Chapman & Hall/CRC, FL, 1993)
37. K. Rummukainen, Monte carlo simulation methods (University of Oulu, Oulu, 2008). http://www.helsinki.fi/~ummukai/lectures/montecarlo_oulu
38. ATLAS Collaboration Collaboration, G. Aad et al. Observation of a new particle in the search for the standard model higgs boson with the ATLAS detector at the LHC. Phys. Lett. **B716**, 1–29 (2012). http://xxx.lanl.gov/abs/1207.7214
39. J. Ellis, J. Espinosa, G. Giudice, A. Hoecker, A. Riotto, The probable fate of the standard model, Phys. Lett. **B679**, 369–375 (2009). http://xxx.lanl.gov/abs/0906.0954
40. J. Elias-Miro, J.R. Espinosa, G.F. Giudice, G. Isidori, A. Riotto, et al. Higgs mass implications on the stability of the electroweak vacuum. Phys. Lett. **B709**, 222–228 (2012). http://xxx.lanl.gov/abs/1112.3022
41. D.G. Figueroa, Imprints of the standard model in the sky: gravitational waves from the decay of the Higgs after inflation. http://xxx.lanl.gov/abs/1402.1345

Chapter 4
Concluding Remarks

Lack of comfort means we are on the threshold of new insights.
—Lawrence M. Krauss

In this thesis, I have demonstrated that gravitational waves can be used as powerful probes of the early universe. I focussed on two separate topics, tensor perturbations from inflation within a quantum gravity formalism, and gravitational waves from preheating in the presence of a light scalar field.

Chapter 2 showed that using the Ashtekar variables in Cosmology, which are an alternative description of gravitational degrees of freedom, lead to a chiral power spectrum of tensor perturbations. This would have an effect on the TB correlator, making it non-zero and potentially measurable, depending on the strength of the parity violation. Although the BICEP2 collaboration has recently detected B-modes [1] of potentially primordial origin, there is not yet sufficient data to explore the TB correlator in detail. To do this, a full sky analysis is needed, which would enable us to constrain the possible chirality of gravity.

If the fairly large value of $r = 0.2^{+0.07}_{-0.05}$ seen by BICEP2 is confirmed by other experiments, the model of massless preheating studied in Chap. 3 has become more viable again. Naturally, this value of r will probably change as more data becomes available, but the observation has undoubtedly given a boost to the simpler models where inflation happens at high energy scales. I have shown that during massless preheating, a light scalar field with superhorizon fluctuations would result in an anisotropic GW background today. Although we cannot currently measure this background, in the future such aniosotropies might provide a vital clue as to how the preheating process occurred, and give further constraints on inflationary models.

The study of B-mode polarization of the CMB marks the beginning of our exploration of primordial gravitational wave backgrounds. Current and future B-mode experiments [1–3] should be able to enhance our understanding of these tensor perturbations, and hopefully one day we might also be able to detect cosmological gravitational wave backgrounds directly. Additionally, detectors like LIGO should

© Springer International Publishing Switzerland 2015

L.B. Bethke, *Exploring the Early Universe with Gravitational Waves*,
Springer Theses, DOI 10.1007/978-3-319-17449-5_4

be able to measure gravitational waves emitted by astrophysical sources in the next few years, which would provide us with fascinating new insights into the world of Astrophysics [4].

Gravitational waves were first predicted by Einstein in 1916, and since then much effort has been invested in understanding their production in the universe and their significance for Astrophysics and Cosmology. Now, nearly a hundred years later, we can finally begin to properly explore these ripples in spacetime.

References

1. BICEP2 Collaboration Collaboration, P.A.R. Ade et al., BICEP2 I: detection of B-mode polarization at degree angular scales, http://xxx.lanl.gov/abs/1403.3985
2. J. Errard, The new generation CMB B-mode polarization experiment: POLARBEAR, http://xxx.lanl.gov/abs/1011.0763
3. B. Crill, P. Ade, E. Battistelli, S. Benton, R. Bihary, et al., SPIDER: a balloon-borne large-scale CMB polarimeter, http://xxx.lanl.gov/abs/0807.1548
4. B. Sathyaprakash, B. Schutz, Physics, astrophysics and cosmology with gravitational waves. Living Rev. Rel. **12** (2009) 2, http://xxx.lanl.gov/abs/0903.0338

Appendix

A.1 Cosmological Perturbation Theory

In this appendix I want to describe the basic features of cosmological perturbation theory. In particular, I will discuss the SVT decomposition into scalars, vectors and tensors and the importance of choosing a gauge.

When we define a metric in GR, we need to choose a threading, which corresponds to timelike lines through spacetime (fixed $\mathbf{x}$), and a slicing, corresponding to spacelike hypersurfaces (fixed t) [1]. For an FRW background, Eq. (1.12), there exists a preferred coordinate system, with a threading according to comoving observers (which measure zero momentum density) and an orthogonal slicing of homogeneous hypersurfaces [2]. However, as soon as we define perturbations, there is no obvious choice of coordinates, and you could even pick a threading and slicing such that it looks like the spacetime is unperturbed [3]. Selecting a specific coordinate system is referred to as picking a gauge [1], and the choice of gauge strongly depends on the problem you want to solve.

The most general form of a perturbed FRW metric can be written as [2]

$$ds^2 = -(1+2\Phi)dt^2 + 2a(t)B_i dt dx^i + a^2(t)\left[(1-2\Psi)\delta_{ij} + 2E_{ij}\right]dx^i dx^j \, . \tag{A.1}$$

Φ is called the lapse (which relates coordinate and proper time [1]), B_i the shift (which measures the relative velocity between the threading and worldlines orthogonal to the slicing [1]), Ψ the spatial curvature perturbation and E_{ij} (which is traceless) the shear.

The energy-momentum tensor, Eq. (1.6), also needs to be perturbed. The energy density ρ and pressure P determine the background FRW metric, so they need to be supplemented by perturbations $\delta\rho$, δP (which depend on space and time). As we are not considering a perfect fluid anymore, we also need a momentum density p_i and a traceless and symmetric anisotropic stress tensor Π_{ij} (which are both zero to zeroth order). For details on the exact form of the perturbed stress energy tensor see e.g. appendix A of [2].

© Springer International Publishing Switzerland 2015

L.B. Bethke, *Exploring the Early Universe with Gravitational Waves*,
Springer Theses, DOI 10.1007/978-3-319-17449-5

A.1.1 SVT Decomposition

To study the relationship between the metric and matter perturbations, we need to expand Einstein's equation (1.1) to first order. This will give us evolution and constraint equations [1]. Note that perturbation components that transform as vectors (B_i, p_i) can be further decomposed into the divergence of a scalar and a divergenceless vector, and similarly objects transforming as tensors (E_{ij}, Π_{ij}) can be decomposed into scalar, vector and tensor parts [2].

Due to the symmetry of the FRW background, scalar, vector and tensor perturbations decouple and all evolve independently (having different symmetry properties under rotations) [1]. This is called the SVT decomposition and greatly simplifies the calculation. We will focus on scalar and tensor perturbations as the vector components are not sourced by inflation [2] and furthermore decay with the expansion of the universe [4]. We therefore care about four scalar perturbations Φ, B, E, Ψ, sourced by the scalar stress energy perturbations $\delta\rho$, δP, p, Π, and a transverse and traceless tensor perturbation, which we will call h_{ij}^{TT}, solely sourced by the transverse traceless anisotropic stress, Π_{ij}^{TT}. Tensors therefore only couple to matter distributions that have a non-zero anisotropic stress [4]. This is not the case for the inflaton, so they are not directly sourced by inflation.

A.1.2 Gauge Selection

Due to the diffeomorphism invariance of GR, we can always make a gauge transformation $x^\mu \rightarrow \tilde{x}^\mu = x^\mu + \xi^\mu$, i.e. a first order change in coordinates, which leaves the form of the metric invariant [5]. To find out how such a transformation affects different types of perturbation, we need to invoke the tensor transformation law (1.2). Scalars do not transform, but they do shift their position, so the new set of metric perturbations $\tilde{\Phi}$, $\tilde{B}$, $\tilde{E}$, $\tilde{\Psi}$ will be linear combinations of the old ones. The perturbations in the stress energy tensor also transform. We find that there are two redundant degrees of freedom in the metric perturbations that can be eliminated by appropriately picking two scalars ξ_0, $\xi_{,j}$ that determine the gauge transformation [4]. The tensor perturbation h_{ij}^{TT}, being transverse ($h_{ij,i}^{\mathrm{TT}} = 0$) and traceless ($h_i^{i\,\mathrm{TT}} = 0$), is gauge invariant [4].

Selecting an appropriate gauge is particularly important for the scalar perturbations, which are the source of matter density perturbations. In this case, it is useful to define gauge invariant quantities, either just in terms of metric perturbations, or for combinations of both stress energy and metric components [2], like for example the curvature perturbation $\mathcal{R}$ in Eq. (1.68). This makes it easy to relate quantities defined in different gauges. As scalar perturbations are not the main focus of this thesis, I refer the reader to [1, 2, 4] for details on popular gauges in Cosmology.

A.2 Hamiltonian Constrained Systems

Constrained Hamiltonian systems and their quantization have been extensively studied by Dirac [6, 7]. All gauge theories (like, for example, GR, where the local symmetry transformations that leave the theory invariant are the diffeomorphisms [8]) need to be supplemented by constraints in their Hamiltonian formulation [9], which take care of the fact that the theory should not change under symmetry transformations. In this appendix I will summarise the most important aspects of Dirac's procedure, in particular focussing on the meaning of primary, secondary, first and second class constraints.

First, let us recall the basics of Hamiltonian mechanics [10]: Starting from a Lagrangian $L(q, \dot{q})$, we can derive canonical momenta $p = \frac{\partial L}{\partial \dot{q}}$ and define the Hamiltonian by performing a Legendre transform,

$$H(p, q) = \dot{q} p - L \,. \tag{A.2}$$

Hamilton's equations

$$\dot{q} = \frac{\partial H}{\partial p}, \quad \dot{p} = -\frac{\partial H}{\partial q}, \tag{A.3}$$

are equivalent to the Euler-Lagrange equations and can be expressed in terms of Poisson brackets (which become commutation relations upon quantization). The Poisson bracket is defined as

$$\{f, g\} = \frac{\partial f}{\partial q}\frac{\partial g}{\partial p} - \frac{\partial f}{\partial p}\frac{\partial g}{\partial q}, \tag{A.4}$$

and Eq. (A.3) can therefore be written as

$$\dot{q} = \{q, H\}, \quad \dot{p} = \{p, H\} \,. \tag{A.5}$$

The time evolution of any function $f(q, p)$ of the canonical variables can similarly be expressed as the Poisson bracket with the Hamiltonian,

$$\dot{f}(q, p) = \{f, H\} \,. \tag{A.6}$$

Note that the Hamiltonian is supposed to be expressed only in terms of q and p which relies upon the fact that the Lagrangian is non-singular, such that the relation $p = \frac{\partial L}{\partial \dot{q}}$ can be inverted to find the velocity $\dot{q}$ in terms of the canonical momentum p. This is not possible in general and the reason why the standard Hamiltonian procedure needs to be generalised. The Lagrangian is always singular in the case of gauge theories [9].

If the Lagrangian cannot be inverted, it means that that the phase space variables are related, i.e. they satisfy a constraint [7]

$$\phi(q, p) \approx 0 \,. \tag{A.7}$$

The use of the approximately equal $\approx$ signals that this is a weak equality, i.e. one that is only satisfied after the equations of motion have been imposed. In particular, we can only impose this condition after Poisson brackets have been evaluated [7]. There may be several such conditions $\phi_m(q, p) \approx 0$ on the phase space variables and they are known as the primary constraints.

We can take care of these constraints by extending the Hamiltonian to

$$H_T = H + u_m \phi_m , \tag{A.8}$$

where u_m are arbitrary coefficients (Lagrange multipliers) that do not depend on (q, p). Clearly $H_T \approx H$, so the extended Hamiltonian reduces to the ordinary Hamiltonian when the equations of motion and therefore the constraints are satisfied. Now, we can easily incorporate the constraints by deriving the equations of motion for any function $f(q, p)$ using the generalised total Hamiltonian H_T in Eq. (A.6).

We need to satisfy $\dot{\phi}_m \approx 0$ to ensure that the constraints are conserved. Eq. (A.6) gives us consistency relations for each of the primary constraints. Using the Hamilton equations (A.3) to solve (A.6) for ϕ_m leads to three separate cases [7]: For some values of m, the consistency condition is identically satisfied, giving no new conditions. In other cases, we might obtain further, secondary constraints $\phi_k(q, p) \approx 0$, which also need to be conserved, i.e. plugged into (A.6). This means that overall we actually have to satisfy $\dot{\phi}_j \approx 0$, $j = 1...M + K$ where there are M primary and K secondary constraints. Lastly, we might obtain conditions that enable us to uniquely determine some of the coefficients u_m, while others remain undetermined. As I will describe now, the latter play an important role for gauge theories.

The primary constraints $\tilde{\phi}_a$ (which in general are a linear combination of the original primary constraints ϕ_m) that correspond to the undetermined Lagrange multipliers u_a are first class [7]. A variable R is first class if it satisfies the condition

$$\{\phi_j, R\} \approx 0 \quad \forall j , \tag{A.9}$$

i.e. the Poisson bracket with all primary and secondary constraints is zero. It is straightforward to show that the total Hamiltonian is first class, and so are the $\tilde{\phi}_a$ [7]. The first class primary constraints are very important as they are generating functions of infinitesimal transformations that preserve the physical state. In other worlds, they generate gauge transformations.

Heuristically, the reason the $\tilde{\phi}_a$ generate gauge transformations is that the undetermined, arbitrary Lagrange multipliers u_a reflect the fact that the phase space variables (q, p) cannot be determined uniquely from an initial state. However, they need to correspond to the same physical system regardless of the value of u_a, in the same way that an arbitrary gauge transformation needs to leave the theory invariant [7].

Constraints not satisfying Eq. (A.9) are called second class. They correspond to redundant physical degrees of freedom, and can be taken care of by using the Dirac bracket [6] instead of the Poisson bracket: it gives the same time evolution as before, but makes it possible to set the second class constraints to zero [7].

In canonical GR, all the constraints are first class and satisfy the symmetries of general covariance: At each point in space there is a 3d diffeomorphism constraint, corresponding to diffeomorphisms on spacelike slices, and the Hamiltonian constraint, corresponding to time translations [11]. Hence, there seems to be no meaningful way to describe the evolution of a system with time in the context of canonical GR. In the Ashtekar formalism, the formulation in terms of a complex spin connection means there is also an SU(2) gauge group, which leads to an additional Gauss constraint [12].

Note that in the quantum theory, the constraints are not imposed on the operators, but as conditions on the Hilbert space: physical states need to be annihilated by the constraints [12]. This is what gives rise to the Wheeler-DeWitt equation [13] in canonical quantum gravity, $\mathcal{H}|\Psi\rangle = 0$, where $\mathcal{H}$ is the quantum Hamiltonian constraint.

References

1. A.R. Liddle, D.H. Lyth, *Cosmological Inflation and Large-Scale Structure*. (Cambridge University Press, Cambridge, 2000)
2. D. Baumann, TASI lectures on inflation, http://xxx.lanl.gov/abs/0907.5424 0907.5424
3. V. Mukhanov, *Physical Foundations of Cosmology* (Cambridge University Press, Cambridge, 2005)
4. S. Dodelson, *Modern Cosmology* (Academic Press, Elsevier, 2003)
5. J.A. Peacock, *Cosmological Physics* (Cambridge University Press, Cambridge, 2001)
6. P.A. Dirac, Generalized Hamiltonian dynamics. Can. J. Math. **2**, 129–148 (1950)
7. P. Dirac, *Lectures on Quantum Mechanics* (Dover Publications Inc, New York, 2001)
8. R.M. Wald, *General Relativity* (The University of Chicago Press, Chicago, 1984)
9. A.W. Wipf, Hamilton's formalism for systems with constraints, http://xxx.lanl.gov/abs/hep-th/9312078hep-th/9312078
10. L.N. Hand, J.D. Finch, *Analytical Mechanics* (Cambridge University Press, Cambridge, 1998)
11. G. Date, Lectures on constrained systems, http://xxx.lanl.gov/abs/1010.2062 1010.2062
12. P. Dona, S. Speziale, Introductory lectures to loop quantum gravity, http://xxx.lanl.gov/abs/1007.0402 1007.0402
13. B.S. DeWitt, Quantum theory of gravity. 1. the canonical theory. Phys. Rev. **160**, 1113–1148 (1967)

CPSIA information can be obtained at www.ICGtesting.com
Printed in the USA
LVOW02*1739040615

441217LV00001B/4/P

9 783319 174488